Frederic H. Wagner

A Chanticleer Press Edition

WILDLIFE OF THE DESERTS

BOOK CLUB ASSOCIATES, LONDON

First frontispiece. *A harvester ant (Veromessor) carries the flowering head of a plant to its subterranean nest in the Sonoran Desert. The nest's granary may contain more than a liter of seeds, stored for use in dry years.*

Second frontispiece. *Although quite unrelated in their evolutionary history, Australian thorny devils (Moloch horridus; shown here) and North American horned lizards (Phrynosoma) are similar in their appearance and ant-eating habits. Such organisms are called "ecological equivalents."*

Third frontispiece. *Gemsbok (Oryx gazella) run across dunes of the Kalahari Desert in southern Africa. While many African hoofed mammals prefer areas ranging in vegetation from grasslands to forests, these arid-land creatures prefer dunes and open terrain. Gemsbok form nomadic herds of 30 to 40 animals.*

Overleaf. *Dunes of the Namib Desert in southwestern Africa stand in sweeping, stark contrast behind a pond of fresh water. Though austere and abrasive, sand is a porous medium through which rare moisture can percolate down to plant roots or flow into the subsurface water table to produce oases.*

Library of Congress Cataloging in Publication Data

Wagner, Frederic H
Wildlife of the deserts.

(Wildlife habitat series)
"A Chanticleer Press edition."
1. Desert biology. I. Title. II. Series.
QH88.W29   574.90954   79-24137
ISBN: 0-7112-0024-6

Library of Congress Catalog Card Number: 79-24137

This edition published 1980 by:
BOOK CLUB ASSOCIATES
by arrangement with Windward

Windward
An imprint owned by W.H. Smith and Son Limited
Registered number: 237811 England
Trading as:
WHS DISTRIBUTORS
Euston Street
Freemen's Common
Aylestone Road
Leicester, LE 2 7SS

Composition by Dix Typesetting Co. Inc., Syracuse, New York
Printed and bound by Dai Nippon Printing Co., Ltd., Japan

Note: Illustrations are numbered according to the pages on which they appear.

Prepared and produced by Chanticleer Press, Inc., New York:
Publisher: Paul Steiner
Editor-in-Chief: Milton Rugoff
Managing Editor: Gudrun Buettner
Project Editor: Mary Suffudy
Assistant Editor: Mary Beth Brewer
Production: Dean Gibson
Art Associate: Carol Nehring
Picture Librarian: Joan Lynch
Map: Herbert Borst, Francis & Shaw, Inc.
Design: Massimo Vignelli

General consultants: Eric Pianka, Professor of Zoology, The University of Texas; John Farrand, Jr., The American Museum of Natural History

Appendixes, text: pp. 196–197, 200–201, and 204–207, by Allen Rokach, The New York Botanical Garden; pp. 198–199, 202–203, and 216–221, by Eric Pianka; pp. 208–215, by Richard Spellenberg, New Mexico State University

Appendixes, art: pp. 194–197 and 200–215, by Paul Singer; pp. 198–199 and 216–221, by Dolores Santoliquido

Contents

Preface

For many people, the word "desert" conjures up an image of lifeless sand dunes stretching off to the horizon and shimmering in midday waves of heat. Dictionaries define deserts as barren, uninhabited regions; to be "deserted" is to be abandoned or forsaken. And yet to many, visiting a desert inspires the same wonder and has the same spiritual impact as contemplating the ocean or a mountain has on others. To these "xerophiles," the awesome vistas, cloudless skies, and dazzling sun afford the same sense of belonging to an ultimate reality greater than themselves as does the sight of the heavens on a starry night.

Regardless of what the dictionaries say, deserts are far from uninhabited. No desert is without some forms of life. To be sure, the very driest regions—those with less than 50 millimeters of rainfall a year—are sparsely inhabited; but more than three-fourths of what are commonly called deserts have annual precipitation ranging from 50 to 250 millimeters and generally have very diverse plant and animal life. Because living conditions are severe, desert plants and animals must possess highly specialized characteristics of structure and behavior in order to survive. As a result, they are particularly interesting to professional ecologists and amateur naturalists alike.

All the major deserts of the world are inhabited by people, and have been since well back in prehistory. Far from feeling abandoned or forsaken, these people look on the desert as their home. They understand the sources of sustenance it contains as well as its hazards. It becomes a part of their pastimes, their art, their religion. Indeed, many desert dwellers have considered the city a source of evil and looked down on sedentary city types. To this day, many individuals who have moved to the city because of its cultural advantages or the need to make a living, escape for periods of time to open country and live like nomads in tents.

What we normally think of as deserts cover about 15 percent of the Earth's land area, and one person out of every 20 in the world is sustained by their biological productivity. This may seem like a comparatively small fraction of the world's population, and it may be thought that the food and fiber needs of such beneficiaries of the desert could be provided from other parts of the world. Hence, loss of the desert's productivity through improper use and inadequate conservation may not seem to be any great cause for concern. But it *is* cause for concern. The world's population is continuing to grow at about 2 percent a year, which means that in the next 30 to 40 years it will double. With half or more of the world's people now inadequately fed, clothed, and housed, we will need to extract twice as much food and fiber from the Earth's surface in the next half century just to keep our standard of living from declining. To lose any of the Earth's existing productivity would make an already staggering problem all the more difficult.

Furthermore, the world's deserts are now misused and are producing at less than their potential. With proper management, they could yield more than they do now to relieve the crowding and deprivation that prevails in many regions.

A given area of the Earth may contain dozens of species of plants and vertebrate animals, several hundred species of invertebrate animals, and countless microscopic plants and animals. All these plants and animals, climate and soils, and their myriad interactions are called an *ecosystem*. Such a system functions as a unit, which over a period of decades or centuries remains intact and experiences only short-term fluctuations.

To some, understanding such a complex ecosystem as a desert may seem beyond the time and capabilities of the average person. But, in fact, the innumerable interactions proceed according to a limited number of general principles or natural laws. It is no more of a challenge to learn these principles and processes than it is to learn the regulations and tactical nuances of soccer or the harmonic and orchestral aspects of a symphony.

With that understanding, each type of habitat on Earth takes on new meaning and worth. Whether desert or lake or tropical forest, each has its own integrity and beauty. To allow any part of it to deteriorate into a truly barren waste means not only to squander some of the productivity that sustains mankind but also to despoil a beauty that raises our quality of life above bare survival.

The emphasis of this book is on the wild animals of desert ecosystems, because such animals attract the most attention and are therefore most helpful in developing an ecological understanding. But that interest need not be merely passive or sedentary. Increasing numbers of people are going into the field to learn about wild animals; and as they do so, they find that animals have both practical as well as aesthetic and recreational value. That is not to say that all animals are useful to all desert peoples. For the hunter-gleaner cultures of the Kalahari, Australian, and western American deserts, animals provide food, clothing, shelter, and implements. But for the pastoral or agricultural societies of North African and Asian deserts, certain desert animals may represent a threat to livestock or crops.

An understanding of animal ecology may also help us protect the environment. Many animals, particularly predatory forms, are sensitive indicators of ecological change. In a sense they are an early-warning system—a signal that conservation measures must be taken before it is too late.

For such diverse reasons, the study of desert animals can be rewarding and valuable. This book is intended to make those rewards and values more widely available.

My sincere thanks go to several people who helped so much in the development of this book. Marilyn Wagner conducted library searches, gave valued critical advice, and typed the various drafts of the manuscript. Milton Rugoff, Mary Suffudy, Mary Beth Brewer, and others at Chanticleer Press provided the much-needed editorial work, photographs, and illustrations and helped in shaping the final result.

Frederic H. Wagner

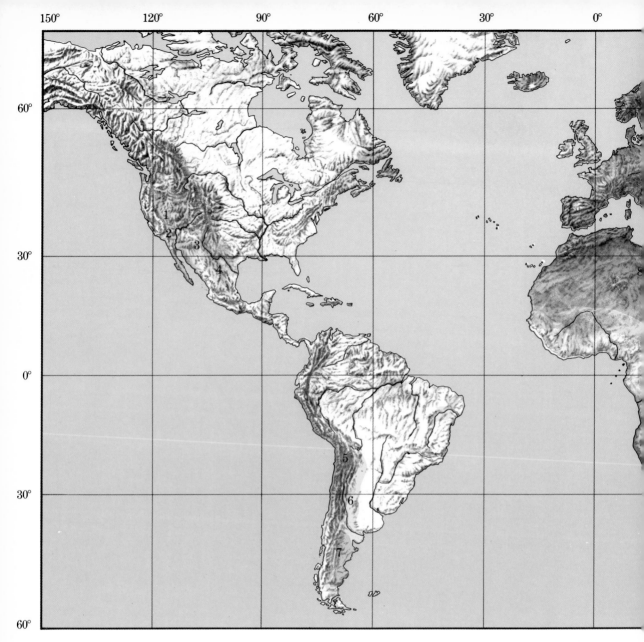

The darker-colored areas on the map represent those regions of the world considered desertic or arid. (These areas cover about 15 percent of the Earth's land surface.) Lighter-colored areas are those generally classified as semiarid. Arid conditions occur in broad bands centered approximately over 30° north and south latitude. They also occur on the downwind sides of mountain ranges in areas called rain shadows, and in downwind continental interiors. Two unique, especially arid, but cool and cloudy deserts occur along the west coasts of southern Africa and South America, where the above conditions are combined with proximity to cold ocean currents.

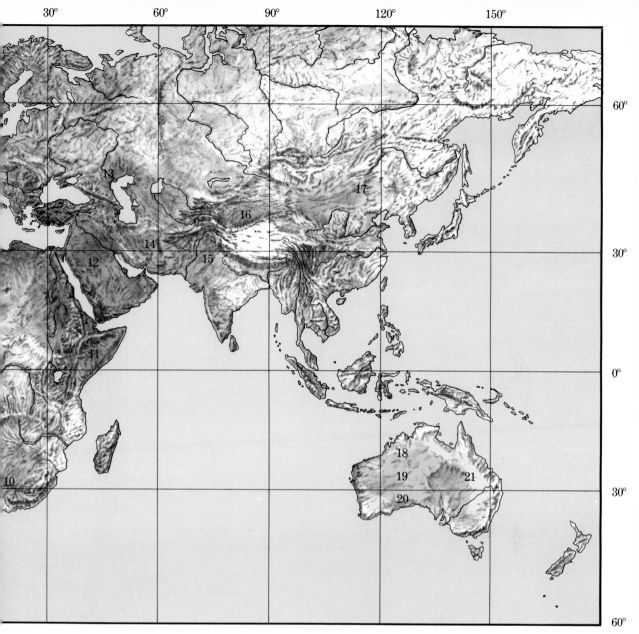

1. Great Basin
2. Mojave
3. Sonoran
4. Chihuahuan
5. Atacama-Sechura
6. Monte
7. Patagonian
8. Sahara
9. Namib
10. Kalahari
11. Somali-Chalbi
12. Arabian
13. Turkestan
14. Iranian
15. Thar
16. Takla Makan
17. Gobi
18. Great Sandy
19. Gibson
20. Great Victoria
21. Simpson

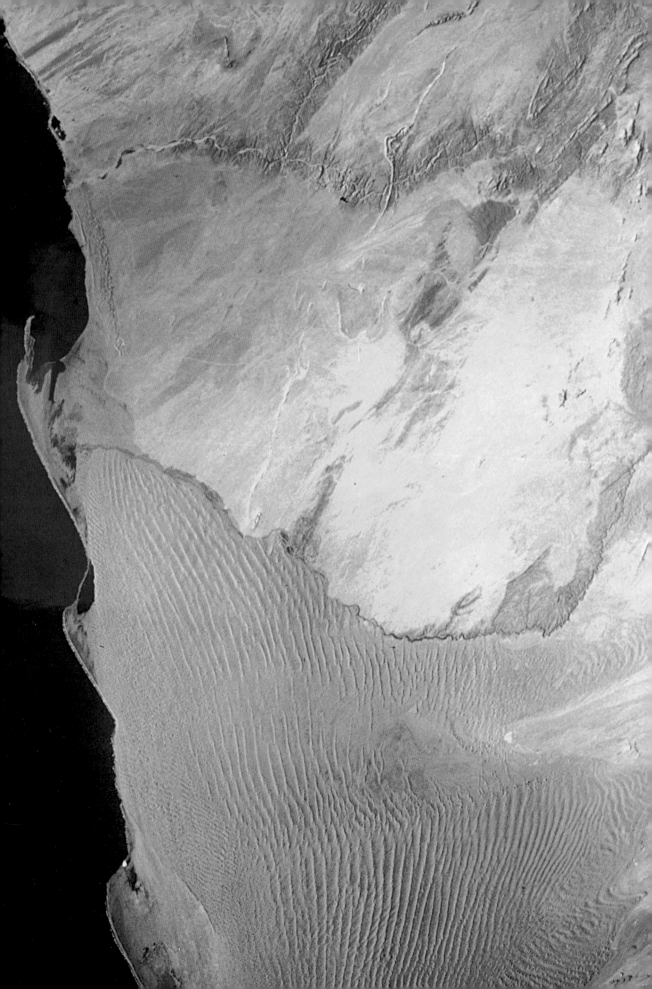

Deserts: The Physical Environment

The conditions in which desert plants and animals must evolve and survive are demanding: blazing sun so intense that it drives air temperatures in the shade up to 50°C; windblown sand so abrasive that it carves rock into bizarre sculpture; powder-dry soil that has received no rain for ten years. The physical environment determines what plants and animals will reside there. In forest and grassland, the vegetation can mute and modify the physical environment to a degree; but plant life in deserts is so sparse that it can play only the slightest role in buffering heat, wind, and aridity.

There is considerable variation in the nature of deserts around the world, and ecologists are not in full agreement about precisely what constitutes a desert. In general, deserts have hot temperatures for at least part of the year, and they are dry. Usually, deserts occur in areas where annual precipitation does not exceed 250 millimeters, and where, if water were present in an open container, more would evaporate each year than falls as rain. Although the polar regions have low rainfall, their low temperatures disqualify them as typical deserts.

Air Currents That Create Deserts

Satellite photographs clearly reveal a definite patterning of the globe's vegetation, with which most people are vaguely familiar. Because of the very close link between vegetation and climate, the polar regions are covered with ice caps, whereas tropical forests develop toward equatorial latitudes. None of the types of vegetation in a locale occurs at random; rather, the climate, topography, and soil of a region dictate vegetation. To understand desert ecosystems, we must take into account where the climates that create deserts occur and the conditions that produce those climates.

At the equator, the Earth's surface is nearly perpendicular to the sun. But as the Earth curves away from the equator, it also curves away from the energy source, so that the ground surface at the poles is virtually parallel to rays from the sun. Hence the surface of the Earth is heated much more at the lower latitudes than at the higher. As the temperature of the surface rises, it heats up the surrounding air. Since surface temperatures are highest near the equator, air temperatures are also highest there. The density of air is a function of its temperature: the higher the temperature, the lighter the air. Since surface air at the equator is the warmest—and therefore the lightest—in the Earth's atmosphere, equatorial air tends to rise. Polar air, on the other hand, tends to descend. Because gravity holds the atmosphere around the Earth like a film, the rising air at the equator does not flow out into space; rather, it tends to accumulate over the equator and is forced to spread out, moving toward the poles at high altitudes. At the poles, the descending air also cannot accumulate indefinitely near the ground and so moves toward the equator, to fill the void left by rising air, which meanwhile fills the void above the poles.

If the Earth did not rotate on its axis, this is the pattern we would see. There would be two zones or belts of *descending* air, one at each pole, and a third belt of *rising* air, at the equator. And we would see two mammoth cells of circulation enveloping each hemisphere, with air rising at the equator, flowing aloft toward the poles, and

14. *The Namib Desert and the southwest coast of Africa are visible in a Gemini V space satellite photograph. The Kuiseb River separates pink dunes to the south from rocky desert to the north.*

16–17. *Rainfall along the coastal portions of the Namib Desert is so scant that perennial plants cannot survive. But in those unusual years when there is rain, annual grasses sprout, flower, and set seed. In the next few years, until the rains come again, the dried grasses provide food for a community of detritivorous beetles and other insects, which in turn feed spiders, lizards, moles, snakes, and birds.*

Opposite. *Without any vegetation to break the wind's force, small soil particles are blown away and leave behind only a rocky surface, such as that of Sturts Stony Desert in Australia. These surfaces are called "gibber" in Australia, "gobi" in Asia, "reg" in the Middle East and North Africa, and "desert pavement" in North America.*

returning to the equator near the Earth's surface. We would perceive this surface flow as north winds in the Northern Hemisphere and southerly winds in the southern half of the globe.

But it's not that simple, for this pattern is broken up by the Earth's rotation. At the equator, the Earth rotates on its axis from west to east at about 1,700 kilometers per hour; in higher latitudes, the velocity declines until it is zero at the poles. The atmosphere rotates along with the Earth's surface. But the Earth tends to slip from under the atmosphere, and the "slippage" is greatest at low latitudes and declines toward the poles.

Thus, the Earth is ringed with six, rather than two, great cells of atmospheric circulation that determine the global climatic pattern and, in turn, the distribution of plant and animal life. This complex circulation pattern is made up of three belts of *rising* air: one at the equator and one each at about 60° north and 60° south. Rising air cools as it ascends into the lower-temperature strata aloft. And since cold air cannot hold as much water vapor as warm air, cooling condenses moisture in the air and produces rain in these belts of ascending air masses. Thus the equatorial region and the zones around 60° north and south latitude have heavy precipitation.

There are four belts of *descending* air around the Earth: one around each pole and one each at about 20° to 30° north and 20° to 30° south. Since descending air becomes warm and can hold more moisture, it not only fails to give up rain but actually absorbs moisture from the Earth's surface. As a result, the polar areas have relatively little precipitation and are sometimes called *polar deserts*.

The true deserts of the world—the Sahara (the most extensive in the world), Australian deserts, North American deserts, Middle Eastern and Indian deserts, and the southern African and South American warm deserts—occur mostly in the two warm subtropical belts of descending air. While these subtropical belts produce the most extensive areas of aridity, certain land features and surface airflow cause arid conditions in other less extensive regions. Again, rather than the two cells of atmospheric circulation we would expect to find on a stationary Earth, there are in reality six cells sandwiched between seven belts of ascent and descent.

In the Northern Hemisphere there is a cell of rising air at 60°, with a flow aloft toward the North Pole, descent around the pole, and a southerly return near the surface between the pole and 60°. A second cell encompasses rising air at 60°, flow aloft toward 30°, descent around 30°, and a northerly surface return between 30° and 60°. A third cell presents a pattern of rising air at the equator, flow aloft to 30°, descent there, and a southerly surface return to the equator. The Southern Hemisphere has mirror images of these patterns.

The change in velocity of the Earth's rotation from the equator to the poles not only forms these seven belts of ascent and descent, and the six great cells of circulation, but also deflects the direction of surface airflow within the cells. Between the equator and 30° north, the surface flow toward the equator moves from northeast to southwest; these winds are called *trade winds*. Between 30° and 60° north, the surface flow toward the pole moves from southwest to northeast; these winds are called *westerlies*. North

of 60° north, surface flow away from the pole moves from northeast to southwest; this airflow is called the *polar winds*. A mirror-image airflow pattern occurs in the Southern Hemisphere.

These surface winds pick up moisture over the oceans and deposit it as precipitation on the continental landmasses. The western edges of North America and Europe above 35° to 40° north have relatively damp climates because the westerly winds blowing off the Pacific and Atlantic oceans carry a great deal of moisture to the land.

As these winds sweep across landmasses, they rise and cool; their moisture condenses and falls as precipitation. Consequently, as these winds progressively lose their moisture, continental interiors tend to be drier than continental margins. Although the entire arid and semiarid region of Australia—some two-thirds of the continent—lies within the subtropical high-pressure zone, the driest part of the continent is its center. The Gobi Desert of Mongolia, like moist Western Europe, lies between about 40° and 45° north latitude, in the zone of westerlies. But by the time the westerlies reach the Gobi, in the eastern portion of Eurasia, they have blown across most of the largest landmass on earth and no longer carry moisture.

Rain Shadows

If prevailing winds are forced to rise over mountain ranges, they cool sharply and lose their moisture more quickly. Along the western edge of the United States, two mountain ranges run from north to south: the lower Coastal Range and the higher Cascade and Sierra Nevada ranges. Moisture-laden westerlies sweeping inland from the Pacific are forced to rise over these ranges, in the process cooling and losing their moisture on the mountains' western slopes. Bereft of moisture, the winds descend the eastern slopes of the Cascades and Sierra Nevadas, drying the land in their path. The result is the Great Basin Desert of the western United States.

Such leeward sides of mountain ranges lie within what are called *rain shadows*. The Great Dividing Range along the east coast of Australia blocks Pacific moisture carried by the southern trade winds and places most of the continent in a rain shadow. The Takla Makan Desert of Sinkiang Province in China is surrounded on three sides by mountains. The Patagonian Desert in South America is in the rain shadow of the Andes.

Dry Air and Reflected Heat

While daytime humidities in summer commonly range around 80 to 90 percent in forested regions, they are typically around 10 to 20 percent in desert regions, and may rise no higher than 20 to 40 percent at night. In humid regions, a large proportion of the sun's energy is reflected or absorbed by clouds and water vapor in the air. As a result, less than half of the incoming radiation may reach the ground. But in deserts, cloudless skies and the near-absence of moisture in the air offer almost no barrier to incoming radiation, most of which strikes the ground. For this reason desert regions receive more solar energy than the tropics, despite their higher latitudes. Desert air temperatures in the shade commonly reach 40–45°C, sometimes as high as 50°; but the ground surface may heat to temperatures as high as 70–80°C.

Above. *When the silty soils of internal drainage basins (playas) dry out, they crack into intricate mosaics, as on the Kuiseb Delta in Namibia* (top) *and the Orange River bed* (bottom), *separating South Africa and Namibia.*

Opposite. *Playas are characteristic of deserts the world over. Rare rains run off the landscape, collect in the basins, and evaporate. The playa shown here is in Nevada.*

Overleaf. *The Atacama-Sechura Desert of South America is a coastal desert like the Namib. Dunes—such as the one shown here near San Juan, Peru—form across the prevailing winds, with the gradual slope on the windward side and their steep, slip face on the lee side.*

Any object that absorbs radiant energy also radiates it. As an area on the Earth's surface rotates away from exposure to the sun during the night, it gives up more energy than it receives, and its temperature falls. In humid regions, clouds and humidity act as a partial barrier to this reradiation, just as they have blocked part of the incoming energy during the day, so that the temperature falls no more than 5–10°C at night.

In the desert, however, the moisture-free air provides no significant barrier to nighttime reradiation, just as it has blocked very little incoming energy during the day. Hence, at night, temperatures fall as much as 20–25°C, and deserts thus have a great daily range in temperature.

How Desert Land Features Are Formed

One of the most fascinating aspects of desert regions is the scenery. In their environs the titanic forces of nature combine to produce some of the most spectacular landforms to be seen anywhere on earth. While some observers find the barrenness and lack of vegetation stark and inhospitable, it is easy to see that dense vegetation would mask much of the rugged and fascinating topography and cloak the reds, golds, browns, blacks, and grays of the bedrock and soils that make desert scenery so colorful.

In general, large topographic features on the surface of the Earth are formed by two constantly repeated, opposing forces: those which raise the Earth's crust into mountains, ridges, and plateaus, and those which wear these landforms down again, namely, the erosive action of water, wind, and ice. Water alone has a wearing action, but if it also carries a suspended load of sand or gravel as it runs off, its abrasive ability is greatly magnified. Wind, too, can carve the landscape, especially when it is laden with sand.

In general, erosion proceeds more rapidly in humid areas because greater rainfall subjects the land to more abrasion. As rainwater flows over the mantle of vegetation in a humid area, it often takes on an acid cutting strength from tannin in the foliage. In arid areas, where rainfall is infrequent, erosion takes considerably longer. Streams arising in more humid regions or high atop mountain ranges can, in their flow across arid regions, slice gorges through plateaus and leave behind flat-topped mesas and buttes that may persist for millions of years. The spectacular canyonlands of the southwestern United States could not have endured long in a humid climate.

Erosion in arid regions produces some characteristic land features—mountain ranges with their intervening basins; buttes and mesas with their intervening valleys and gorges. Since rising air cools and loses moisture, mountaintops in desert areas get more rainfall than the surrounding lowlands. The resulting runoff cuts ravines and canyons in the mountain walls and carries along such erosion material as soil, gravel, and boulders, which is deposited in a fan of alluvium radiating from the canyon mouth. As numerous canyons are cut along the face of a mountain range, their fans coalesce to form around the base of the range a continuous alluvial skirt, known in North America by the Spanish word *bajada*.

When runoff from a mountain range reaches the base and begins to slow down, it first deposits the boulders and large rocks. As the flow spreads out over the valley floor

Above. *At the northern end of Death Valley National Monument in California, the rigid contours of Ubehebe Crater are being slowly worn away by wind, sand, and weathering. The eroded material forms an alluvial skirt.*

Overleaf. *At Bryce Canyon National Park in Utah, wind and water erosion of the sandstone stratum is in the early stages. With little vegetation to mask them, exposed desert soils and geological strata display varied red, gold, black, and gray hues.*

Above, opposite. *Many of the world's deserts, such as that of Petrified Forest National Monument in Arizona, are geologically young. These logs are remnants of a formerly luxuriant forest where desert now occurs. Color variations of petrified wood come from mineral solutions of iron and manganese that once penetrated the logs.*

and slackens further, the sand and silt settle out of the water.

Consequently the alluvium at the upper edge of the bajada may be quite coarse, while that farther down in the basin may be quite fine clay or loam. One can see zones of vegetation along a bajada slope that reflect this zonation of alluvial particles. Often the most diverse vegetation along a bajada slope occurs near the top, where larger boulders, rocks, and soil particles offer plants a greater variety of places to grow.

Another feature of desert landscapes is dry streambeds, called *washes* or *arroyos* in North America, *oeds* in North Africa, and *wadis* in the Middle East. They may arise in a mountain range and wind across the floor of the desert. They remain dry except after rare rain showers, when they collect runoff and carry it over the landscape. Wadi banks usually have more vegetation than the rest of the desert. Viewed from the air, many deserts reveal a generally thin sprinkling of plants, with the wadi network represented by thin lines of slightly more dense vegetation.

The valley floors between desert mountain ranges, called *playas* in North America, *chotts* in North Africa, and *qas* in the Middle East, sometimes collect runoff waters from the wadis to become shallow lake beds. A playa may hold water for a few weeks at most, during and after the rainy season. As soon as the runoff stops, the collected water evaporates. But in areas collecting runoff from more extensive regions, especially from large mountain ranges, the inflow of water may be continuous and offset the evaporative loss. In these cases, playas have year-round standing water and become lakes, such as the great chotts of southern Tunisia, the Great Salt Lake of North America, and the Dead Sea of Israel and Jordan.

Desert playas and lakes are usually salty because runoff water contains a number of dissolved chemical elements, especially sodium and chlorine, the ingredients of table salt. When water evaporates from a playa, these chemicals are left behind and, year by year, the mineral content of a playa increases. The salt concentration of the Great Salt Lake and the Dead Sea is about eight times the average salinity of the oceans.

Desert Soils

For many people, the word "desert" evokes endless vistas of sand rippled by dunes; and indeed, there are vast areas of sand dunes in the world's deserts. The Grand Ergs of the Sahara are commonly called "sand-dune seas." Other extensive dunal areas occur in Saudi Arabia, the Gobi Desert, the Thar Desert of western India, the Namib Desert of southern Africa, and the Great Sandy Desert of Australia. But the total area of stabilized sand, heavier soils, and small rocks over the world's deserts exceeds bare sand and dunes. Since rain is scant in desert regions, desert soils retain their mineral nutrients. Rainwater often reaches perhaps 20 to 30 centimeters below the surface. Once the skies clear, the moisture begins to evaporate in the hot sun. As the ground surface dries, the water beneath is drawn upward and also gradually evaporates. This upward movement of water in the desert's shallow soil layers brings minerals to the upper soil layers and actually deposits some on the surface. Desert soils are often salty, especially in small depressions that catch puddles of water,

where thin white crusts of salt gradually form on the ground surface.

Desert soils may be well endowed with mineral nutrients, but they are extremely poor in organic matter, since vegetation is sparse in desert regions and very little decayed plant material goes into the soil. Nevertheless, desert soils' high mineral content is reasonably good for agriculture where their texture is not too rocky and where water is accessible. With irrigation, huge arid areas in the United States, Mexico, Libya, Soviet Asia, Israel, Saudi Arabia, and India have become very productive.

The scarcity of vegetation in deserts means that the soil has less protection from wind and water erosion. Surface particles are often blown away, thus exposing the rock and pebble substratum, until the ground surface eventually is completely covered with rocks. These desert pavements are called *regs* or *hamada* in North Africa and the Middle East, and *gibber plains* in Australia.

Where the soil particles blown out from around the hamada are sand, they pile up into dunes, which essentially are moving waves of sand that form at right angles to the prevailing winds. The long slope of a dune faces into the wind, whereas its lee side drops off more abruptly. A dune appears to move when the sand is blown up the windward incline and over its crest.

Soil and Moisture

Plants draw water from the soil through their roots. The size of soil particles affects this extraction of moisture and how the vegetation flourishes. Fine soil tends to resist infiltration of water more than does coarse soil. Given a 10-millimeter rain, some of the water will puddle and merely run over the surface of a fine clay soil. Furthermore, moisture that does enter the soil may not filter down more than 10 to 15 centimeters. A sandy soil may absorb all the rainfall to a depth of 40 or 50 centimeters.

Coarse-grained soil also favors plant growth because it gives up its water to the plant roots with greater ease. Soil water is either free liquid, which moves through soil because of gravity or evaporative attraction from the surface, or microscopically fine films bound tightly around each soil particle. The free water, available to all plants having roots deep enough to reach it, is used up first. Once this is exhausted, the plants must turn to the particle films of moisture. Because these films cling much more tenaciously to fine particles than to coarse ones, plants are able to extract more of the bound water present in sandy or rocky soils.

This property may be one of the reasons why desert vegetation tends to be more varied, more attractive, and often more abundant on the bajadas, where soils are coarse, rather than in valley bottoms, where they tend to be fine clays. Conditions for plant growth are harshest in the valley bottoms, where only a few species that are most successfully adapted to drought can survive.

We now have some idea of the nonliving environment in which desert organisms exist. On plants and animals alike, the desert imposes its harsh regimen of high temperatures and scarce water. Its conditions determine the nature of the vegetation in which animals must seek the food and shelter necessary for survival.

Opposite. *After the waters of Lake Manley in Death Valley evaporated, only thin pinnacles of salt remained. Fresh water contains tiny amounts of salt, and when enough of it evaporates in desert regions, lakes like the Great Salt Lake and the Dead Sea are left behind. Eventually, dry salt covers the ground.*

Overleaf. *Frost cloaks the shrubs of Monument Valley, in the western United States. Isolated buttes and spires represent the final stages of erosion. The erosion cycle proceeds swiftly in areas of high rainfall, where frequent runoff erodes and levels land features. But where rain is scant, spectacular landforms persist for long periods, providing arid-land visual feasts.*

Desert Life-styles

Among the miracles of nature are the spectacle of an evening primrose on a vast sand dune, a rabbit hopping across a gravelly desert wilderness, a lizard slithering along a stony expanse—all exposed to a withering sun, with no water and little or no food anywhere around them. How do such living things manage to survive and flourish? Since the desert environment is one of the most severe on Earth, desert organisms have developed some of the most extreme adaptations to cope with the hardships they face. As a result, they offer especially vivid examples of natural selection, the culling out of unfit individuals.

Among the survival problems facing arid-land creatures are getting and conserving water, avoiding overheating, getting enough food, and avoiding attack by predators. However, solutions to these problems leave desert plants and animals between the horns of several dilemmas, for in solving one problem the organism may increase its susceptibility in another area.

Plants face a special dilemma. They manufacture the sugars they use for food by combining water, carbon dioxide from the air, and energy from sunlight. When plants open their pores to let in carbon dioxide, however, they lose water in a process called *transpiration.* In areas of copious rainfall, plants can compensate for this loss by taking up water from the soil as fast as they transpire it. But in deserts the dry air exerts a strong drying force, and the plant is hard put to make up for this loss out of powder-dry soil. Desert animals have a similar problem. As they forage for food in dry air, they too lose water through their skins, especially during the heat of the day. Like the plants, they face a perilous balance between starvation and desiccation.

The risks of overheating present a second dilemma. Plants must expose their leaves to the sun in order to absorb the light, but in doing so, their temperatures rise, potentially to the point of tissue damage. In areas of ample rain, water lost in transpiration causes evaporative cooling; but in the desert, water needed for such a cooling process is scarce. Animals active in sunlight also take up heat. While they can cool off by releasing moisture, as many do by sweating or panting, that release brings the all-too-familiar risk. In addition, by growing or foraging in the open, plants and animals expose themselves to attack by other animals. The traits or adaptations that enable desert plants and animals to walk the thin line between these dilemmas are the distinguishing characteristics of these organisms.

Getting Water

The nurseryman's rule of thumb in transplanting an ornamental tree or shrub is that the plant has about as much tissue below ground as it has leaf and branch tissue above. With this amount of roots, the plant can extract soil water fast enough to offset transpirational loss.

Desert plants commonly have two to six times as much tissue below ground as above. Some desert trees and shrubs send out large, woody roots laterally through the soil for 50 to 75 meters until they reach a wadi or arroyo. The roots then descend into the wadi bed and get moisture derived from the broad land area drained by the wadi. This heavy commitment of growth to roots at the expense of leaves and stems explains in part why the aboveground portions of desert plants are small.

34. *At an oasis in the Atacama-Sechura sands near Ica, Peru, the water is bitter and contains no living organisms.*

36–37. *Burchell's zebras (Equus burchelli) drink at a rare water hole in southern Africa. Zebras are plains animals, but they venture into the desert when rain occurs there.*

Above. *Streams that cross deserts provide habitats and migration routes for aquatic species from more humid regions. Although this frog* (Rana catesbeiana) *is a common species in eastern North America, this individual is basking in the west-flowing Colorado River, the headwaters of which interdigitate with streams that flow east.*

Left. *This painted turtle* (Chrysemys) *lolls in a Chihuahuan Desert spring pond near the headwaters of the east-flowing Rio Grande. The long claws of this animal identify its male gender.*

Desert animals get their water in several ways. The smaller forms such as insects, reptiles, birds, and rodents, which make up most of the desert fauna, cannot travel great distances to scarce water holes and must obtain moisture from their food. Water constitutes roughly half the content of seeds, stems, and old leaves, and even more of fruits, young foliage, and the tissues of succulent plants such as cacti. By weight, animal bodies are two-thirds to three-fourths water, and in turn they supply the moisture for the predators that feed on them.

Although most desert animals do not need free water, they will drink it when available, as after a rare rain or if a water hole is within reach. The abundance of life around desert water holes shows that many animals thrive better with a supply of free water, even though it is not absolutely essential to their survival. Water holes are centers of activity for insects, birds, mammals, and reptiles that all gather to drink. Predators take advantage of this behavior and lie in wait or hunt in the vicinity. Many species wait until dark for their visits to a water hole, evidently to avoid such daylight-hunting predators as hawks and eagles.

Conserving Scarce Moisture

It is in their adaptations for conserving moisture that desert organisms are most distinctive. Plants have evolved a number of ways to minimize transpirational loss. One is to reduce their transpiring surfaces. Many have only tiny leaves and rely on their stems, which have few pores, to take over much of the foodmaking role. The desert broom (*Retama raetam*) of North Africa and the Middle East and the paloverdes (*Cercidium*) of North and South America are plants of this kind. In other desert plants that have no leaves at all, photosynthesis takes place entirely in the green stems: for example, the *Ephedra* of the Kara Kum Desert in Soviet Asia and the *Casuarina* of the Australian desert.

Some plants grow leaves after a rain, when there is enough soil moisture, and photosynthesize at that time. Then, as soils dry out, they shed their leaves and await the next rain. These are called *drought-deciduous* plants, and the North American ocotillo (*Fouquieria splendens*) is a classic example. An ocotillo may grow several generations of leaves within a single year or else, in especially dry years, none at all.

A characteristic of desert vegetation is the wide spacing between plants. This is thought to result from the competition among plant roots for moisture. The drier the area, the more widely spaced are the plants. There is evidence that some desert plants actually secrete toxic substances in their leaves or roots which get into the soil and prevent young plants from becoming established close by—a sort of chemical warfare against competitors.

Some desert plants have evolved the ability to store water after rainfall. The cacti and yuccas of North and South America, aloes of South Africa, and euphorbias of North Africa and Asia are called *succulents* because they develop thick, spongy leaves and stems for absorbing and holding large quantities of water. Moisture taken up by their roots after rain is gradually used in dry periods.

Succulents have an ingenious method for minimizing transpirational loss. Photosynthesis and metabolism are, in a sense, mirror images of each other. In photosynthesis, a

41 top. *River gums* (Eucalyptus) *grow in a dry riverbed in Western Australia. Like many of the 600 species of* Eucalyptus *in Australia, these have white bark that reflects sunlight and prevents excessive heating.*
Center. *A tree appears to grow miraculously out of a quartz ledge in the Namib Desert.*
Bottom. *Southern African aloes such as this kokerboom tree lily* (Aloe dichotoma) *have evolved succulent life forms—ability to store water in leaves and stems—as have African and Asian euphorbias and Western Hemisphere cacti and yuccas.*

Overleaf. *Rainbow bee-eaters* (Merops ornatus) *of Australia are aerial insect hunters that nest in holes in sandy banks.*

The towering blossom of a century plant (Agave shawii) in Baja California, Mexico, portends its demise. The plant will sprout from seed and grow leaves for 20 to 50 years before sending up a flowering stalk. The year after it blooms, the plant dies. Contrary to its name and popular legend, it does not grow a full century before blooming. Fibers of the century plant make excellent ropes, used by Mexican cowboys (vaqueros). When fermented, the juice of this succulent makes the intoxicating drink pulque, which when refined becomes tequila.

plant combines carbon dioxide, water, and energy from sunlight to make sugars, and oxygen is expelled as a by-product. In metabolism, a plant or animal breaks down the sugar-like substances in its food, combines them with oxygen it takes in, and releases for its body functions the energy originally incorporated by the plants. The waste products of metabolism are carbon dioxide and water. Thus, the waste products of one reaction are the building blocks of the other.

Succulents have evolved the ability to hold the carbon dioxide they release in metabolism and then reuse it in photosynthesis, thereby actually recycling their own carbon. Of course, a succulent would never grow if it merely reused in photosynthesis the same carbon released in its metabolism. It must add carbon dioxide from the air, so that it fixes more carbon in photosynthesis than it releases in metabolism. But by storing and recycling its carbon, it can keep its pores closed until nighttime, when relative humidity is higher, and until after a rain, when it can offset transpirational loss.

In a remarkable evolutionary parallel, some desert animals also recycle metabolic by-products in order to conserve moisture. But in their case, it is the metabolic water that is retained and recycled by the kidneys rather than being passed out as waste. By recycling metabolic water and by reducing the water content of its urine, the American kangaroo rat (*Dipodomys*) expels only minute amounts of urine in a day.

Such water-conserving methods and efficient kidney function have evolved in unrelated desert animals. They occur not only in kangaroo rats but also in the African and Asian jerboas (*Jaculus*), the Australian spinifex hopping mouse (*Notomys alexis*), the Old World gazelles (*Gazella*), the New World pronghorn antelope (*Antilocapra americana*), and Australian kangaroos. This development of similar traits in unrelated species in response to similar environments is a prime example of *convergent evolution*.

The Long Siesta

By far the greatest number of desert plant species avoid the risk of drying out by having no active stems, leaves, or flowers aboveground during dry periods. Annual plants, such as the carrots and marigolds of our gardens, live and die in a single growing season, leaving behind only their seeds as the living organisms to carry over to the next year. Annuals like Sturt's desert pea (*Clianthus formosus*) of the Australian deserts, the *Coreopsis* of the North American Mojave Desert, and the Iranian *Linaria michauxii* are among the most numerous of desert plants. Desert soils are full of their seeds, waiting for enough rainfall to cause them to germinate. Once it rains sufficiently, their seedlings sprout and develop roots, stems, leaves, flowers, and a new generation of seeds all within a few weeks. The growth sequence must be fast to enable completion of their life cycle during the short period when there is enough soil moisture.

There is a risk in the annuals' life-style. If they germinated after too little rainfall, they could start growth only to have the soil dry out before their life cycle was complete. They might then die before setting their seeds. In deserts, the annuals have special mechanisms to protect against this risk. Their seeds will germinate only after rains substantial

The variety of plant forms in the desert may be second only to those of the tropics. Annuals, perennial ephemerals, and succulents such as the examples shown here are among the common forms adapted for existence where rainfall is scant and unpredictable.
Top row, left. *South African succulent.* Center. Calochoritus kennedyi, *Arizona.* Right. Langloisia punctata, *Death Valley.*
Bottom row, left. Clianthus formosus, *Australia.* Center. Stapelia hirsuta, *South Africa.* Right. *Desert flower of Argentina.* Bottom. *Compositae, Mojave Desert.*

Overleaf. *The saguaro cactus (Cereus giganteus) of the North American Sonoran Desert may grow to 15 meters in height, weigh as much as 7,000 kilograms (5,000 of this being water), and live to be 200 years old. It sometimes grows less than a centimeter a year and does not send out its first branches until it is 75 years old. Its fruits are edible.*

enough to recharge the soil moisture sufficiently to ensure completion of the life cycle. One desert botanist believes that there are chemicals in the seed coats of desert annuals which inhibit germination; only after enough moisture has bathed the seeds to wash out these inhibitors can they germinate.

Many desert insects that behave in the same way are, in essence, "annual animals." Their eggs may live in the soil for several years, waiting for enough rain to allow them to hatch and complete a life cycle. Such plants and animals are called *ephemerals*, and their unpredictable appearance makes desert ecosystems extremely variable.

Plants that retain living branches or roots from one year to the next, even though they may shed their leaves, are called *perennials*. Among desert perennials, a number of species survive drought periods as underground, carrot-like roots or onion-like bulbs. These fleshy roots and bulbs store moisture and plant food, which support their metabolism while underground and the early stages of their growth after the onset of the rainy season. Once stems and leaves develop aboveground, the plant manufactures more food to enable it to grow further and flower and to replenish the supplies in the root that will be needed for the next dry period. As the dry period sets in, the aboveground parts of the plant die, and it reverts to its subterannean sleep.

Lovely examples of these plants are found in the Negev Desert of the Middle East. There, one finds the desert tulip (*Tulipa amplyophylla*) and Negev iris (*Iris atrofusca*) gracing the desert in spring after the winter rains, much as the less hardy, domesticated tulips and irises grace our spring gardens.

Such vertebrates as birds, mammals, and lizards and some invertebrates such as spiders and scorpions are the perennial animals of the desert, and some of them spend the protracted hot, dry periods underground in an inactive state. In mammals and birds this condition is called *torpor*. Many desert rodents become torpid in the middle of summer. Torpor is the desert equivalent of hibernation in cold climates.

The Danger of Overheating

Since plants cannot move, they cannot avoid absorbing heat by getting out of the sun. So some have other means of avoiding temperature rises that would damage their tissues or kill them. One is to reflect the sunlight that strikes them. Many desert plants, such as *Atriplex halimus* of the Negev Desert, grow fine, silky white hairs on their leaf surfaces, which give them a pale, silvery cast and reflect sunlight. Others such as the *Eucalyptus* of Australia secrete a whitish waxy coating on their leaves that serves the same purpose. This coating may effectively reflect more than half of the sunlight striking their leaves. Many desert animals evolve similarly, their reflecting mechanism being pale feathers or fur.

Another way in which some desert plants compensate for inability to move out of the sun is to present as little surface as possible to the sun during the hottest part of the day. Cacti and euphorbias point their stems upward into the sun so that only the tips are exposed to direct sunlight. Eucalypts do the reverse, by drooping their leaves toward the ground. As a result, they intercept little sunlight. Animals, too, avoid direct exposure to the desert sun

Two antelope ground squirrels (Spermophilus leucurus) perch on a cholla cactus (Opuntia) in the North American desert, seemingly oblivious to its needle-sharp spines. In winter these animals are sheltered in underground burrows, from which they venture out during spring, summer, and early fall.

Above. *The bannertail*
(Dipodomys spectabilis) *is the
largest of the North American
kangaroo rats. These two
animals, jousting for territorial
dominion, use their tufted tails as
rudders in their ricocheting gait
—and perhaps as decoys for
fooling predators that chase them.*

Left. *The dusky hopping mouse*
(Notomys fuscus) *of Australia,
shown here, is of the same family
as the Old World house mouse*
(Mus musculus) *and Norway rat*
(Rattus norvegicus). *But long
hind legs and bipedal locomotion,
as well as a long tufted tail,
indicate a marked convergence
with the unrelated kangaroo rats*
(Dipodomys) *and Old World
desert jerboas* (Jaculus).

during hot periods, either by seeking the shade of plants, rocks, or ridges or by going underground. Many rodents and lizards move into their burrows during the day. Midday relative humidities in the desert summer commonly fall as low as 10 percent—levels at which most animals would dehydrate. But the humidity in underground burrows remains commonly 30 to 40 percent. At night, when aboveground humidities rise to 25 or 30 percent, the animal can venture forth to find food.

Getting Rid of Heat

Despite the efforts of desert organisms to avoid excessive heat intake, some heating is inevitable in the high desert temperatures. Plants have only limited ability to regulate their temperatures and must endure a wide range. If they can avoid extremes, they survive reasonably well. Among animals, two physiologically different groups can be distinguished. Birds and mammals fall into one group, which has often been called *warm-blooded*, but in modern ecological terminology it is more properly called *endothermic*, meaning "internally heated." These animals maintain their own body temperatures at constant, high levels by releasing large amounts of heat while metabolizing their food. Birds normally have body temperatures of 40–41°C, and mammals maintain theirs at about 37°C. If birds and mammals are exposed to cold environmental temperatures that drain their body heat, they consume more food, increase their metabolic rates, and generate heat at a faster rate. But when external temperatures are high, as in the desert, they face the reverse problem. Although they can lower their metabolic rates to some degree, their continuous high body temperatures indicate the need for releasing some heat if their internal temperature is not to rise to a dangerous level. One way in which mammals release heat is, once again, to stay underground where it is cooler during the day and then venture forth at night. Birds seek shade and remain inactive during the heat of the day.
Another way of releasing heat is to increase the surface area of the body from which the animal can radiate it into the environment. In the course of evolution, some animals have developed thin bodies and necks and long, slender limbs, which give a maximum of body surface relative to its volume. The Dorcas gazelle (*Gazella dorcas*) of Africa and Asia is a small, delicate, beautifully proportioned animal of this type.
Other mammals are able to get enough water from their food and conserve the moisture for evaporative cooling. Zebras and gazelles sweat; foxes, wolves, and many desert birds evaporate water from their lungs and mouths by panting. To promote such cooling evaporation, several species of Australian kangaroos and wallabies moisten their bodies by licking their fur.
The second group of animals, including reptiles and invertebrates, release less heat generated from their food intake and can regulate their body temperature internally to only a limited degree. Hence their body temperature is strongly influenced by the conditions of their environment. These *ectothermic* (externally heated) animals can raise their temperature to or above the surrounding air temperatures by absorbing heat from sunlight, by bringing their bodies in contact with a warm surface, and by taking up heat

Opposite. *The black-tailed jackrabbit* (Lepus californicus) *of North American arid lands is actually a hare, not a rabbit. Hares are born in dished-out areas on the ground surface after six weeks of gestation. At birth they are fully furred, have open eyes, and are able to move almost immediately. Rabbits are born after four weeks; their eyes are closed, and they are naked and helpless. Rabbits spend the first two or three weeks of their lives in fur-lined nest holes.*

Above. *The bat-eared fox* (Otocyon megalotis) *of the southern and eastern African deserts looks like a small jackal. Its ears, like those of the jackrabbit, are heat-radiating organs.*

from warm air. Others can generate some heat by rapid
movement.

When air temperatures fall to low levels—especially on
cold nights, when there is no solar radiation to be absorbed
—body temperatures among this second group fall. At this
point their actions become extremely slowed, even to a
state of complete inactivity when it is quite cold.

But ectotherms must also guard against overheating when
exposed to the desert sun in the warm seasons. Like
endotherms, they resort to such means as seeking shade or
burrowing underground, and many are active at night in
summer. Moreover, they can tolerate a much wider range
of body temperatures than endotherms can.

Staying Fed in an Unpredictable Environment

Above-average rains may occur in deserts during only one
year in five, with little or no rain in two and average rains
in two. Extremely dry deserts may receive no rain for
several years in a row. Furthermore, the precipitation
pattern is highly irregular and unpredictable.

In such an environment, food storage—common among
many desert rodents—appears to be a solution to the
problem of food scarcity. In rainy years, North American
kangaroo rats and pocket mice (*Perognathus*) and African-
Asian jerboas and gerbils (*Gerbillus*) gather seeds and
store them in their burrows, as do seed-eating desert ants.
While rainfall in any one desert location is highly variable,
over a large geographic region it will occur each year in at
least scattered locations. If an animal cannot store food, it
must have the freedom and mobility to go wherever there
is food. Such is the strategy of the nomad, and most of the
large grazing animals and many birds of the world's deserts
are nomadic.

Avoiding Enemies

Some desert plants have great powers of regrowth after
being grazed upon. Others such as cacti and acacias grow a
formidable armor of thorns to discourage would-be eaters.
Still others develop acrid-smelling or -tasting resins and
terpines in their tissues.

Most animals do not have the restorative powers that
plants have and hence depend more on escape and concealment to evade enemies. In the open terrain of deserts,
special powers of locomotion develop to allow escape.
Numerous desert birds prefer running on the ground to
flying. Over evolutionary time, many lizards and mammals
have come to rise up on their hind legs and run swiftly with
bipedal locomotion. Its most novel form is the bounding
gait of kangaroos; but such a gait has also developed in
many rodent species of the world's deserts and gives the
kangaroo rats of the New World their name.

If an animal cannot flee from its enemies, it may escape
predators by hiding. Many desert species assume pale skin,
fur, or feather colors which blend in with their sun-
drenched environments, thus making them difficult for a
predator to see. (Paleness also helps reflect the sun's rays,
as we have discussed earlier.) But such protective coloration must often be combined with complete motionlessness
if the animal is not to be seen by a sharp-eyed fox or hawk.
Using underground burrows or burrowing into sand, of
course, also enables an animal to conceal itself. So in the
bare terrain of deserts, life underground benefits animals

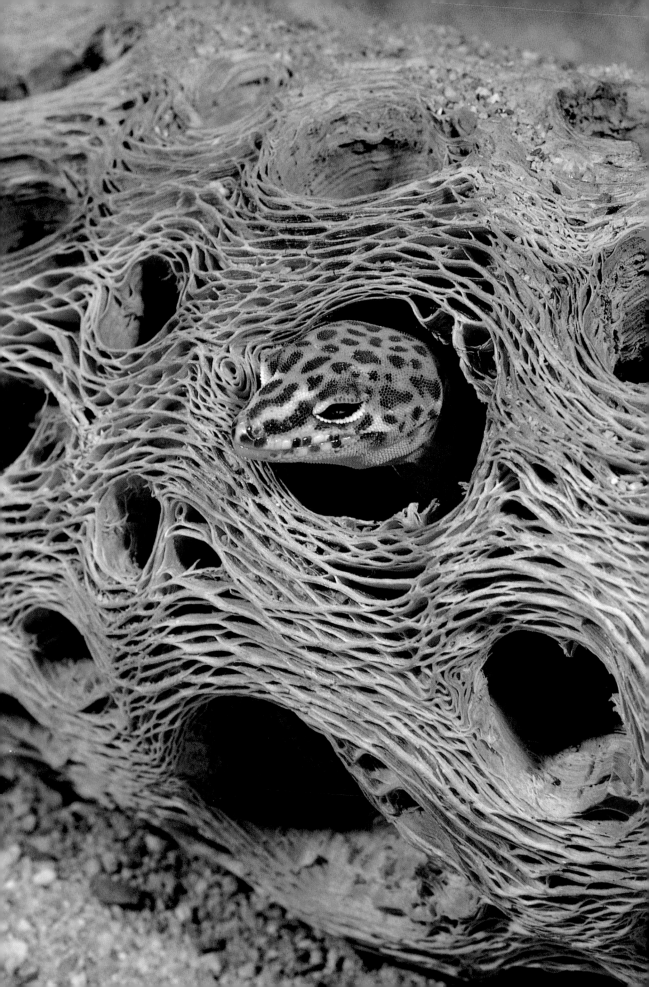

in several ways: conserving water, preventing overheating, and protecting them from enemies.

Divergent and Convergent Evolution

Knowing the adaptations of organisms to their environment is essential to understanding their ecology. But ecologists also want to know how these relationships developed over a period of time, so that they can better understand the end products. For example, in a population of short-eared desert mice, those with slightly longer ears might survive better because they could more effectively avoid too much heat buildup. After a number of generations, the long-eared individuals would predominate in the population. Changes like these do not occur only for a single trait. Body color and shape, leg structure, and behavior patterns, among others, might all change at the same time and eventually produce new species.

Under different environmental pressures, one can imagine other lines of development from the original short-eared mouse. Some might develop larger size and become rat-like; others might develop long tails and an ability to climb trees. Still others might develop a powerful digging capability and assume a burrowing life-style. All would still be rodents, but of very different species.

These diverging lines of evolution are called simply *divergence*. Until recently, most of the emphasis in evolutionary study has been on divergence. But one might ask whether, under the selective pressures of desert environments, short-eared rat-like rodents that entered the desert might over a long period of time evolve long ears like their long-eared mouse cousins. Might the more distantly related tree inhabitants and burrowing life-forms be similarly adapted if they too ventured into the desert? In such a situation, all these species would *converge* in their evolution and develop long ears. Might the selective pressures of deserts be so similar around the world, and the process of natural selection so effective in shaping them, that quite unrelated organisms could converge until they were similar in appearance?

The answer is yes. We have seen numerous cases of convergence in body shape, patterns, of locomotion, leaf size, succulence, kidney function, and other traits. Today the student of evolution traces the patterns of both divergence *and* convergence, each reinforcing the message of the other about effectiveness of natural selection.

There is as much convergence in ecological function as in body structure. In each desert habitat, there is an array of seed-eating rodents, ants, and birds that take advantage of the unique importance of seeds in deserts. A variety of bees that pollinate flowers and darkling beetles that consume dead plant material are also characteristically abundant in most deserts. Such organisms of different ecosystems that have been shaped by similar environments and, even more important, have a similar appearance are called *ecological equivalents*. Jerboas or kangaroo rats, sidewinder adders or sidewinder rattlesnakes, all contribute to the similarities of deserts. This is not to say that there are no differences. A kangaroo is not a gazelle, and a succulent euphorbia is not a cactus. But this combination of variety and unity, this repeated variation on common themes, keeps the professional ecologist enthralled with the study of desert ecosystems.

Opposite. *A* nocturnal banded gecko (Coleonyx variegatus) *peers out a window in the woody skeleton of a cholla cactus* (Opuntia) *in Arizona.*

Overleaf. *A family of burrowing owls* (Athene cunicularia)—*the female is second from the left— surveys the desert landscape from outside their nest burrow in the Chihuahuan Desert of the United States and Mexico. These long-legged birds, which stand 23 to 28 centimeters high, feed primarily on rodents.*

Staying Cool in African and Arabian Deserts

Nowhere on earth is there as large an area with such dry, hot conditions as the Sahara. But this desert has not always been the parched wasteland it is today. There have been wet periods such as that which occurred in Neolithic times, when rainfall was sufficient to cloak the Sahara in enough vegetation to support large numbers of game like those in East Africa today. Prehistoric paintings on rocks and cave walls in the central Sahara depict elephants and giraffes, animals that require large amounts of woody vegetation for food, and hippopotamuses, which spend much of their time in water. In recent years, crocodiles have been found in isolated desert ponds of Algeria watered by runoff from the Atlas Mountains to the north. Their nearest relatives occur in streams in tropical Africa far to the south. Evidently, there were at one time Trans-Saharan waterways running between the tropics and Mediterranean Africa.

Apparently, the climate had changed by the beginning of the Christian Era, for Roman ruins across North Africa do not extend much farther southward than the present borders of the desert. Yet there must still have been abundant wildlife, at least around the desert fringes. The Romans hunted elephants in the steppes of Algeria and Tunisia to send to their circuses in Rome and Carthage. Many of their mosaics in Tunisian museums and at archaeological sites depict scenes of boar, lion, and gazelle hunting.

As late as the 1830's, the early French colonists in North Africa were still reporting extensive herds of gazelles and large numbers of ostriches (*Struthio camelus*). One observer tells of cheetahs (*Acinonyx jubatus*) and leopards (*Panthera pardus*) roaming over much of Algeria—animals that thrive only where an abundance of prey animals is available.

Today those herds are gone from much of North Africa and the Near East. They were first reduced, over thousands of years, by the environmental change from savanna to desert. A smaller number of species that evolved adaptations to the hardships of desert existence still survive today as a legacy of the abundant wildlife resources which can be still be seen in some areas of eastern and southern Africa. More recently, these desert-adapted forms have been reduced to their present remnant status through indiscriminate hunting by humans and the expansion of pastoralism, or animal herding, associated with rapid population growth. Today herds of sheep and goats across North Africa and the Near East and cattle along the southern edge of the Sahara severely overgraze the landscape, leaving little behind for the wild animals.

The Greater Sahara Today

The Sahara proper covers some 10 million square kilometers, an area nearly as large as the continental United States. Although most people think of this great desert as an endless expanse of dunes, in actuality only about one-seventh is sandy, including the major *ergs*, or sand-dune seas. The Great Sand Sea of Libya and Egypt, covering an area the size of France, is the world's largest, with many dunes a hundred or more meters high. Other major ergs occur in southern Tunisia and eastern Algeria, central Algeria, Mauritania, and Niger. Much of the Sahara that is not sandy is covered by reg plains of polished black or

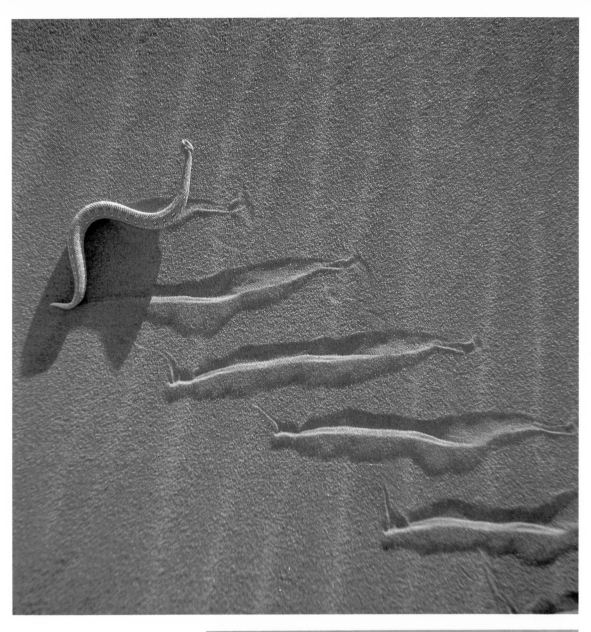

violet stones or by the flat calcareous stones of the hamada.

Topographically, this desert is a region of plateaus and plains crisscrossed by land features that give it, in the eyes of one geographer, a "grandiose simplicity of design." One of its east–west features is the four mountain ranges that march from western Sudan to southern Algeria. Highest of these is the Tibesti Range (up to 3,400 meters), on the border between Libya and Chad. The best-known are the Algerian Ahaggar Mountains (up to 3,000 meters), commonly called the "Saharan Alps." A second major east–west feature is a line of depressions extending from Cairo to southern Algeria. Flanked on the north by cliffs facing south, these depressions accommodate a chain of oases which once provided refuge and sustenance along caravan routes from Thebes to Carthage and which now hold the remains of ancient cities dating back to Pharaonic and Greek times.

The major north–south topographic features are a series of troughs, including the Red Sea, the Nile Valley, and other troughs extending southwestward across Libya and eastern Algeria. The Nile is the only river flowing through the Sahara; the many wadis of other drainages arise in the interior and peripheral mountain ranges and end in interior basins, some at elevations below sea level.

Primarily because of its location astride the subtropical high-pressure latitudes, most of the Sahara has less than 125 millimeters of rainfall yearly. As in all deserts, rainfall is highly variable; in one 30-year period, no rain fell during 17 of the 30 years. Summer daytime temperatures generally rise well into the 40's Centigrade, and commonly exceed 50°C.

The northern third of the Sahara receives moisture primarily between fall and spring. The region's shrubby vegetation provides grazing for Arab sheep- and goat-herders, who were nomadic two generations ago but have since become largely sedentary. The Sahara's central third is the driest zone, with meager, irregular wisps of moisture. Although it has the least vegetation, nomadic Muslim sheep- and goatherders put it to moderate use. In the southern third, called the Sahel, prolonged drought has recently extended the desert considerably. Normally the Sahel receives most of its rainfall in summer. The Somali-Chalbi Desert of Somalia, Kenya, and Ethiopia—an arid strip around the Horn of Africa—is an eastward extension of the Sahel; its summer rainfall also supports grasses that are eaten by cattle.

Although most people think of the Red Sea as the eastern limit of the Sahara, the subtropical high-pressure zone does not stop there but continues eastward across the Arabian Peninsula. The region between Muscat on the Gulf of Oman and Dakar on the Atlantic, further unified by a common Muslim culture, can be thought of as a single vast region, across which the Red Sea is but a thin slice out of an otherwise continuous desert. The Arabian Desert covers 3 million square kilometers, nearly a third of which is sand, the highest proportion of sand in all the deserts of a major continental region. No rivers run into or across it. The most austere extreme of the Arabian Desert is the Rub' al Khali in the southeast corner of the peninsula (known as the "empty quarter"), where dunes may reach heights up to 200 meters.

62. *Cheetahs* (Acinonyx jubatus) *are the fastest mammals on Earth; sprinting across open terrain, they have been clocked at 120 kilometers per hour. Cheetahs commonly feed on gazelles.*

64–65. *Dromedary camels* (Camelus dromedarius), *their nostrils and eyes closed, wait out a Saharan sandstorm in Algeria.*

Above. *A nocturnal web-footed gecko* (Palmatogecko rangei) *emerges from its burrow in the Namib sands at dusk, prepared to venture forth on its nightly foraging.*

66 top. *A sidewinder viper* (Bitis peringuei) *in the Namib Desert moves over sand by thrusting a loop of its entire body laterally and at right angles to its path.*

Bottom. *The fringe-toed lizard* (Acanthodactylus) *finds the Namib sands a hot surface on which to walk. After it takes one step with its left hind and right front feet (raising the other two in the air to cool), it will then tread with the raised feet while cooling the other two.*

68

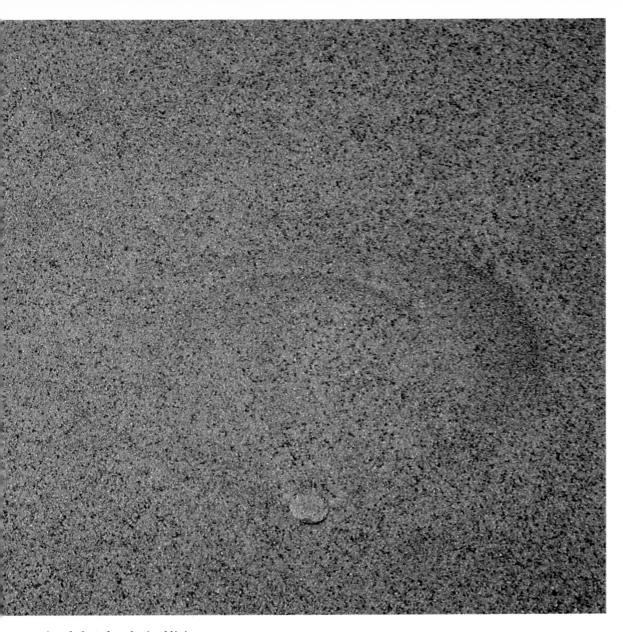

*A web-footed gecko is oblivious
of a sidewinder viper that has
wriggled its body into the sand,
leaving only its eyes and nostrils
protruding. The viper lies in wait
for prey in this manner. The
viper's camouflage is effective, for
it is so well hidden that the gecko
steps on its head.*

Two Deserts of Southern Africa

The equatorial region of heavy rains separates the Sahara from two unusual and less familiar deserts in arid latitudes of the Southern Hemisphere. The larger, with an area of about 600,000 square kilometers, is the Kalahari Desert of Botswana, the Republic of South Africa, and Namibia. It is not as barren as the Sahara and Arabian deserts; its predominantly summer rains produce a sparse, grassy ground cover dotted with trees up to 12 meters high in the north, and with shrubs in the more arid south. Kalahari soils are mostly sandy, often forming parallel ridges that are the remains of ancient sand dunes stabilized by vegetation. Depressions between the ridges catch runoff water during the infrequent rains and form ephemeral water pans. Characteristic of the Kalahari landscape, these pans provide temporary watering sites for much more wildlife than is found in the Sahara. Like much of Africa south of the equator, the Kalahari has not known northern Africa's long history of heavy grazing; hence its vegetation can still sustain more wild animals than other African and Asian deserts.

One of the most unusual deserts of the world is the Namib, a thin strip of land about 100 kilometers wide on the southwest African coast, along the Atlantic coast of Namibia. Although its latitude makes it dry, its location at the point where the cold Benguela Current flows northward out of Antarctic waters, sharply cooling the onshore winds, means that it is chilly and cloaked in fog more than 200 days a year. Despite its dank, cool atmosphere, this driest of all African deserts supports virtually no perennial vegetation. The near absence of ground cover has forced its bizarre fauna—beetles, reptiles, spiders, and other animal groups—to adapt to living in sand.

The Quest for Water

For all desert animals, no matter how varied and ingenious their adaptations to their environment, the quest for water is a vital task. While most get water from the plant or animal tissues of their food or from metabolic water released as they digest the dry parts of their diet, some need free water itself, and others will use it if available. While parts of the Namib have less rainfall than other African deserts, the fog, though lacking sufficient moisture to support perennial plants, does provide drinking water for animal species that have developed ways of condensing it. A number of darkling beetles (Tenebrionidae) engage in fog basking, standing motionless just over the crest of dunes, with their hind legs on the crest and forelegs thrust down the seaward slopes. Thus poised, their backs face into fog moving inland from the coast. The moisture that condenses in tiny droplets on their backs runs down toward their mouths.

The sand vipers (*Bitis peringuei*) gather water in the same manner, coiling on dunes facing the coast. As the fog moves inland, it condenses on their bodies in droplets, which the snakes slowly sip. Only ectothermic, or "cold-blooded," animals such as reptiles and insects can collect moisture in this way, because fog will condense only on surfaces with temperatures lower than that of the air. These animals must radiate heat into space during the night to lower their body temperatures. A continuously warm-blooded bird or mammal could never condense any moisture.

The placement of its eyes and nostrils atop the head enable the sidewinder viper to bury all but these sense organs.

71

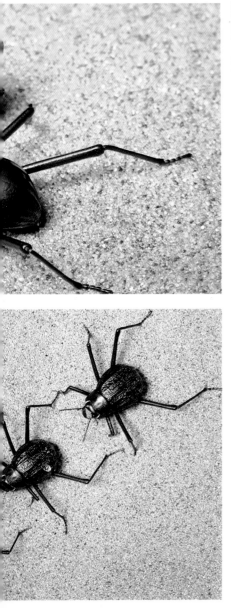

Despite the sterility and abrasiveness of sand, a diverse community of organisms has evolved in the dunes of the Namib Desert.
Top row, left. A darkling beetle (Cardiosis) dives into the sand, to avoid water loss and predators.
Center. Another darkling beetle (Tenebrionidae) eats a wasp (Scoliidae). Darkling beetles are scavengers, or detritivores, which feed on dead plant and animal material.
Right. The scorpion-like solpugid (Metasulpuga picta) of the Namib dunes is a predator that feeds on insects. Solpugids of many species are found in deserts the world over.

Bottom row, left. Shown here is another species of Cardiosis, one of the 200 kinds of darkling beetles peculiar to the Namib Desert. All but the most extreme deserts have diverse invertebrate faunas. It is common to find from several hundred to more than a thousand species in a square kilometer of desert terrain.
Center. Onymacris vegatapennis engage in a mating chase.
Right. The white lady spider will inject sand grains into her web and then drape the web over her burrow. Unsuspecting prey do not distinguish the web from the sand and, venturing over it, become enmeshed.

Bottom. A sacred scarab beetle (Scarabaeidae) in Botswana rolls a dung ball into its hole. It will insert its eggs into the ball, and the hatching young will feed on it.

Many of the Kalahari's water pans may be watering areas that animals themselves created. These pans may have originally appeared near termite mounds, which, because they are constructed of soil cemented with termite secretions, are rich in certain minerals. Large grazing animals lick the mounds just as cows lick salt blocks. In the course of time, sharp hooves continually churning the soil create depressions around the mounds. Wind blows away the pulverized soil, deepending the depressions further, and they eventually become deep enough to catch and hold water after rains. Animals that then come to drink may repeat the process and thus continue to enlarge the pan. Many pans cover several hundred hectares, or even several square kilometers. They draw attractive fauna, including the gemsbok (*Oryx gazella*), zebra (*Equus burchelli*), ostrich (*Struthio camelus*), and springbok (*Antidorcas marsupialis*).

Some birds also require free water daily. One such group is doves, of which there are one or more species in most desert areas. In North Africa, the turtle dove (*Streptopelia turtur*) flies up to 50 or 75 kilometers daily to water holes. Unlike other birds, which drink by scooping up mouthfuls and letting the water trickle back down their throats, doves drink like human beings, by continuously sucking and swallowing. Adult turtle doves feed their newly hatched young on a secretion called "pigeon milk," produced in their crop. When the young bird inserts its mouth into an adult's mouth, the pigeon milk is regurgitated, providing the young with vital moisture. As the young grow, the adult's secretions subside, and the young are fed increasingly on seeds. But since their need for water increases, the adults regurgitate some of the water they get during their daily flights.

The African and Asian sandgrouse have an even more unusual way of carrying water to the young. These species, closely related to doves and pigeons, are also seed-eaters and need free water daily. When the adult sandgrouse fly to water each morning, they immerse their breast feathers, which soak up water like a sponge. When the adults return with this saturated water supply to the nest, the young use their beaks to extract the water from the wet feathers.

Evading Heat Gain

If they are to survive, all desert animals must devise means of avoiding an intolerable increase in body heat and of getting rid of excess heat. One way of avoiding heat buildup is through careful timing of daily activities. A high proportion of desert animals are thus nocturnal and become active in exposed areas only at night, in the late evening, or in the early morning. A walk through the Negev Desert of Israel and Jordan during midday in summer would leave the impression that there are almost no animals in the region. But a walk over that same path at night with a flashlight reveals scorpions and snakes that have emerged from their burrows and are lying in wait for prey, or rodents moving about in search of seeds.

Lizards, birds, and many insects remain active during daylight hours; but during the hottest months, their movements are limited to short intervals at dawn and dusk. In temperate latitudes, most birds begin singing and display about 30 to 45 minutes before sunrise and continue for two or three hours. But desert birds such as the crested

Above. *A male Namaqua sandgrouse* (Pterocles namaqua) *seeks food with its chicks. Like doves and pigeons, to which they are closely related, sandgrouse are seed-eaters. They must have fresh water daily, and will fly 50 to 75 kilometers to get it.*

Left. *Two female yellow-throated sandgrouse* (Pterocles gutturalis) *pause at a water hole in the Kalahari Desert. Sandgrouse have an unusual way of watering their young: their breast feathers are structured so that they absorb water when immersed; the adults then fly to the nest, where the waiting young strip the water out of the soaked feathers.*

lark (*Galerida cristata*), very common across North Africa and in the Near East, are afield a full hour before sunrise, and by the time the sun's rim appears on the horizon their activity has subsided.

Ants, also common desert dwellers, are like other insects, fish, and reptiles: they are ectothermic, and their body temperature is strongly conditioned by their environment; they are vulnerable to overheating and chilling. In late fall and in winter and spring, ants are active at midday when temperatures rise to a comfortable level. But from late spring to early fall, midday temperatures are too high for them, and their activity shifts to morning and evening hours.

Desert animals that are active at night and at dawn and dusk spend their inactive periods in the shade or underground. To those which remain aboveground, shade is precious. In the morning, the desert or Cape hare (*Lepus capensis*) sits under a shrub on the west side, then once the sun has passed its midday zenith, moves to the east side. At that time of peak heat, desert birds remain completely motionless under shrubs or branches. The addax (*Addax nasomaculatus*), an antelope of the Sahara, and the ibex (*Capra ibex*) of the Negev Desert, animals that prefer mountainous and rocky terrain, spend daylight hours in caves or under the shade of rocky overhangs.

For similar reasons, many desert insects, lizards, and rodents adopt a burrowing existence and commonly dig their tunnels in clay or loam at the base of shrubs. Gerbils of African and Asian deserts tunnel a complex system of burrows 1 to 2 meters below ground.

Underground escape is only part of a broader pattern of behavior in which desert animals take advantage of vertical temperature differences both above and below ground. Temperatures, which are the highest at the ground surface, decline abruptly in the first few centimeters above and below ground, then more gradually at greater heights or depths. Desert animals are highly sensitive to such slight differences in temperature. If an animal can climb a shrub only 1 meter above ground, as snakes and lizards commonly do, or can burrow 30 centimeters below ground, it finds an environment 20–40°C cooler than the ground surface.

Even a few millimeters' difference can help an animal avoid the searing ground surface. The ants of one North African species walk with their legs kept as nearly straight as possible, in order to lift their bodies as far from the ground as they can. Furthermore, in order to reduce the ambient temperatures of the abdomen a degree or two, they raise their bodies vertically rather than holding them horizontally as most ants do.

The long legs of the Arabian camel, or dromedary (*Camelus dromedarius*), along with its broad feet, enable it to navigate loose sand with ease. But equally important, those legs raise the camel's body to a level where air temperatures are as much as 25° lower than the ground surface.

Raptors such as the Bateleur eagle (*Terathopius ecaudatus*) of the Kalahari soar effortlessly on updrafts of air rising from the heated ground surface. They glide 300 to 500 meters above ground, where the temperature is 40–50°C lower.

One Saharan bird, the white-crowned black wheatear

(*Oenanthe leucopyga*), evades desert heat by laboring for days to gather several hundred small stones, each one half its own weight, and then stacks them in a pyramid 15 centimeters high. It makes its nest atop this pyramid, thus raising the eggs above the hot desert floor and allowing wind to blow through the porous stack of stones underneath. Its nest is often placed in shade under overhanging cliffs or on the cooler east side of large boulders.

Radiating the Heat Away

Although their strategies help them avoid desert heat, endothermic birds and mammals cannot escape their own internally generated heat, which they must continually expel to avoid overheating. No species show more effective adaptations for radiating heat than the Saharan and Arabian mammals. Slender animals, many with long ears, necks, and limbs, they all have a large body surface area in proportion to their bulk.

The rhim gazelle (*Gazella leptoceros*) of the Sahara is one such animal. Similarly, the camel's long neck, legs, and body, along with the great surface area of the sides of its abdomen, provide a great deal of radiating surface. The long legs and neck of the ostrich serve the same function, and the bare skin of its neck provides a large radiating surface uninsulated by feathers. Many desert species, including the small jerboa (*Jaculus jaculus*) and the desert hare, have long ears. One especially striking example is the tiny Saharan fox, or fennec (*Fennecus zerda*). Smaller than a house cat, this beautiful, pale-colored animal has enormous ears that have greater surface area on one side than is covered by the animal's entire face.

The need to radiate internally generated heat raises another dilemma for desert animals. Because the food supply in the desert is highly undependable, it is to the advantage of a desert animal to form and store fat during periods of plenty, much as succulent plants take up and store water after a rain. But animals usually store fat under the skin, and since fat is good insulation against heat exchange, such storage would make the process of ridding themselves of heat loads more difficult. The solution for desert animals is to concentrate fat stores in a limited area of the body: the camel's hump is a well-known example of this. The fat-tailed gecko (*Pterydactylus dauricus*) of the Jordanian deserts has the same solution.

Evaporative Cooling

Any object loses heat when water evaporates from its surface. A zebra sweats, and so do such other hooved mammals as the gazelle, gemsbok, addax, and ibex. Carnivorous Saharan mammals such as the striped hyena (*Hyaena hyaena*), the black-backed jackal (*Canis mesomelas*), and the sand cat (*Felis margarita*) pant, thereby allowing moisture to evaporate from the tongue and the mouth. While sitting in the shade during the day, desert birds engage in gular fluttering, their equivalent of panting; they open their mouths and vibrate their throats, expelling and drawing in air rapidly. Moisture evaporates off the linings of the lungs.

The Dunes Are Their Home

Loose sand without vegetation is a sterile, abrasive medium. One would think it unlikely that any diversity of

78 top. *Southern carmine bee-eaters* (Merops nubicoides) *gather in the Kalahari Desert.*
Center. *The carmine bee-eater* (Merops nubicus) *has developed a special means of finding prey. It perches on the back of a Kori bustard* (Ardeotis kori) *and rides about until an insect is spotted; it then flies off to catch it.*
Bottom. *The secretary bird* (Sagittarius serpentarius) *is an unusual long-legged, ground-dwelling bird of prey that feeds on snakes and rodents.*

Overleaf. *Ostriches* (Struthio camelus), *the largest birds in the world, run across a dry water pan in Namibia. The ostrich's closest relatives are the South American rhea* (Rhea americana) *and the Australian emu* (Dromaius novae-hollandiae). *Distribution of these flightless birds is one line of evidence that Africa, South America, and Australia were once joined in a huge southern super-continent, Gondwanaland.*

animal life could evolve and live in such an environment. But given millions of years, the evolutionary process has been able to mold organisms of a kind that can survive in most environments on Earth. So it is that the barren dunes of the Namib Desert contain a surprisingly diverse community of animals that have found one way or another to use sand, whether to burrow into it to avoid water loss and the sun's heat, to escape enemies, or to lie concealed waiting for prey. To blend in with the sand, some species have a protective and often beautiful coloration. One species of grasshopper is not only sand-colored but also half-transparent, so that when it remains motionless it is almost invisible.

Other species move through the sand with great ease. After eons of evolutionary selection, a skink lizard (Scincidae) has become legless and can thus "swim" unimpeded through sand, almost as an eel swims through water. The eyes of the skink are also considerably reduced in size. The Cape golden mole (*Chrysochloris asiatica*), perfectly colored to match its environment, has powerful front legs for digging swiftly through the sand to catch its prey, the legless skink. Needing no eyes while burrowing through sand, it has lost these organs and depends entirely on a delicate sense of smell and its other senses.

A number of predatory species have developed striking behavior patterns that help them lie in wait for prey. The sand viper (*Bitis peringuei*) wriggles its body into the sand until only the nostrils and eyes protrude. The eyes are set high on the head so that the entire head can be buried. Coiled and hidden in this fashion, it waits motionless for its prey.

Two species of predatory spiders use sand grains to their advantage. The white lady spider digs a small pit in the sand's surface, then spins a web over the top that incorporates enough sand grains to disguise it. Underneath, the spider lurks in the pit. When its prey, the dune cricket (Gryllidae), walks over the web, it collapses and the cricket drops into the pit.

The back-flip spider also weaves sand grains into a web large enough to cover it. Once the web is finished, the spider flips over on its back and pulls the web over it. The spider then waits for its prey to walk over the camouflaged web and become entangled.

The spider wasp (Pompilidae) subdues its prey by first paralyzing it with its sting; then, using projections along the sides of its legs like those of a garden rake, it digs a burrow into the sand. The wasp then shoves the paralyzed spider into the hole, lays an egg on it, and closes the hole. When the egg hatches, the larval wasp begins feeding on the still-living spider.

Just as the spider wasp has garden-rake legs, the fringe-toed lizard (*Acanthodactylus*) has developed bristles along the sides of its toes that help it walk on the surface of the sand, much as snowshoes serve for walking on snow. The nocturnal web-footed gecko lizard (*Palmatogecko rangei*) has a similar solution; webbing between its toes makes the feet like a buoyant disk.

While deserts certainly do not have the abundance of life found in grassland or forest, they are hardly sterile wastelands. They accommodate a variety of fauna that includes some of the most specialized animals found anywhere.

Namib Desert suricates, or gray meerkats (Suricata suricatta), are small predators related to the mongooses and civets found over much of Africa. They are highly social (living in small warrens) and feed primarily on invertebrates, but occasionally on rodents, small birds, and reptiles. Easily tamed, they have become common as pets.

Above. *African ground squirrels*
(Xerus erythropus) *fight for territory in the Somali-Chalbi Desert
of East Africa. The animals strike
at each other with their hind feet,
but they will not kill or seriously
injure each other.*

Overleaf. *Gemsbok* (Oryx gazella)
and Burchell's zebras (Equus
burchelli) *stand in a dry water pan
in Namibia as a black-backed
jackal* (Canis mesomelas) *strolls
by.*

The Search for Food: Deserts of Central and East Asia

Not until Sven Hedin and other twentieth-century explorers brought back tales of the Takla Makan, Kyzyl Kum, and Gobi deserts and of the vanishing lake of Lop Nor did Westerners have any conception of the arid lands of Central Asia. As other travelers described the rich animal life of the deserts of western India, Occidentals began to realize how vast and varied were the deserts of Asia and how interesting was their animal life.

Central and East Asia have two basic types of deserts. One, represented by the Thar Desert of western India and eastern Pakistan, is a hot, low-latitude desert quite similar to those of North Africa and the Arabian Peninsula; in fact, this is regarded by some as an extension of the latter. The other type, sometimes called "cold desert," extends westward from western Pakistan through Afghanistan and Iran, then curves northward into the southern Soviet Union and back eastward through northern China and southern Mongolia. It lies in a considerably colder latitude than any of the deserts we have considered so far. This climatic difference is less marked in summer, when temperatures there are nearly as high as those of the more southerly deserts, than it is in winter, when temperatures in the northernmost parts of the zone may fall as low as 30–40°C below freezing. These cold temperatures add another dilemma for desert animals: they not only must protect themselves by developing insulation in the form of fur, feather, and fat layers but also must cope with the way that insulation hampers the release of body heat in summer.

As in other deserts, both plant and animal food are less abundant than in grasslands or forests, and the food supply is also extremely undependable, since it varies with the highly capricious rainfall. The severe winters of the cold deserts aggravate these problems. In low-latitude deserts, some insects remain active throughout the winter and provide year-round food for birds and lizards; but since insects are inactive during the northern winter, insect-eaters there face a total absence of food during this season. The same difficulties confront predators that feed on birds and lizards. If there is snow—and these deserts receive much of their limited moisture in winter—food may become even harder to obtain, both for carnivores and herbivores.

Just as they have ways of avoiding overheating and desiccation, desert animals have evolved a number of behavioral, structural, and physiological adaptations that help them get food and endure periods of hardship. Like desert animals elsewhere, many of the animals of the Asian deserts move on to areas where rain has fallen and food is available. Many of the sedentary creatures store food during times of plenty. And because animals come to occupy different parts of the landscape or feed on different kinds of food, they are able to avoid self-defeating competition.

In the Land of the Maharajas

The Thar Desert, situated roughly between 22° and 32° north latitude, is located mostly in western India but extends a short distance into Pakistan. Since prehistory the climate of its 700,000 square kilometers, like that of northern Africa and Asia Minor, has probably become more arid; this change to desert conditions has been aggravated

88. *The Great Central Desert of Iran lies in a basin-and-range physiographic region. Dominant forces shaping the landscape are the continual upward thrusting of the Earth's crustal blocks to form mountain ranges, which are in turn eroded down. Alluvium carved from them is deposited in basins between the ranges, forming the alluvial skirts at the base of mountains characteristic of such regions.*

90–91. *A flight of desert locusts* (Schistocerca gregaria) *in western India virtually darkens the sky. These insects occur from Africa north of the equator, across the Middle East, and into arid regions of Central Asia.*

by intensive human use in the form of livestock grazing and removal of vegetation for fuel and building materials.

As recently as a century or two ago, the predominantly summer rainfall, ranging from 300 millimeters annually at Jodhpur on the east to 150 millimeters along the Pakistan border, supported a savanna-like vegetation with perennial grasses and widely spaced trees. This vegetation fed large numbers of grazing animals such as the Indian gazelle (*Gazella gazella*), the nilgai or blue bull (*Boselaphus tragocamelus*), the blackbuck (*Antilope cervicapra*), the Asiatic wild ass (*Equus hemionus*), and the Indian rhinoceros (*Rhinoceros unicornis*). These herbivores in turn sustained many predatory species, such as the tiger (*Panthera tigris*), lion (*Panthera leo*), and cheetah (*Acinonyx jubatus*). The presence of abandoned hunting lodges, formerly used by maharajas and now surrounded by barren desert, attests to the plentiful wildlife that once roamed the region.

The region also supported many resident birds, whose ranks were swelled in winter by migrants from the north. These, too, attracted hunters; around Bikaner in western Rajasthan Province, parties of British viceroys, governors, and other potentates could in one day shoot as many as 2,000 birds from wintering flocks of imperial sandgrouse (*Pterocles orientalis*). Until recently, wintering flocks of Houbara bustards (*Chlamydotis undulata*) attracted parties of Arab sheikhs from the Middle East, who pursued these birds with trained falcons. By the time the practice was halted in 1978, the bustard population had seriously declined.

Today, the Thar is a barren land, largely devoid of perennial ground vegetation. The tops of the widely spaced *Prosopis cineraria* trees are hedged, their branches lopped off to feed livestock. The soils are sandy, and there are some extensive areas of raw dunes astride the India-Pakistan border. The rhinoceros, wild ass, and large predators are gone, as are nearly all the blackbuck and nilgai. A few gazelles as well as most of the birds may still be found, but in greatly reduced numbers.

The Cold Deserts

The cold deserts include four major regions, each with two to four subdivisions, collectively covering an area nearly half that of the Sahara. The southernmost of these, the Iranian Desert of western Pakistan, Afghanistan, and Iran, covers 500,000 square kilometers. Situated largely between 25° and 38° north latitude, it has a topography of mountain ranges separated by closed basins, with altitudes ranging from 600 to 2,000 meters. Its temperatures, therefore, are distinctly more temperate than those of the Thar Desert and the Arabian Peninsula.

The wildlife of the Iranian Desert is a mixture of species from several deserts—the Thar on the east, the Middle Eastern deserts on the west, and the Soviet deserts on the north—which have been reduced in numbers and variety by long-term climatic changes and severe overuse of the land. Much of Iran is barren as far as the eye can see. Its dunes, up to 250 meters high, are among the tallest in the world.

Situated between the Caspian Sea on the west and the Pamir and Tien Shan Mountains on the east, the Turkestan Desert of the Soviet Union occupies more than 2 million

Above. *Voracious desert locusts devour an ear of millet during a 1968 plague near Keren, Ethiopia.*

93 top. *Seed-eating, metallic wood borer beetles (Buprestidae) mate in the Iranian desert.*
Center. *The praying mantis (Mantidae) feeds on insects, for which it waits motionless on vegetation. Most mantids are cryptically colored.*
Bottom. *A desert weevil, or snout beetle (Curculionidae), feeds on a shrub in the Iranian desert.*

Overleaf. *A stately male black-buck (Antilope cervicapra) strides across the Rann of Kutch, in western India. Once abundant in India's Thar Desert, the species is today nearly extinct.*

These differently eroded strata are in the Khuzistan foothills of the Zagros Mountains in western Iran. The mountain range separates the country's arid interior from the fertile Tigris-Euphrates Valley to the west in Iraq.

square kilometers. The Kara Kum ("black sand") Desert on the southwest and the Kyzyl Kum ("red sand") Desert on the northeast are the major subdivisions of this vast sandy region. East of the Turkestan, in the western Chinese province of Sinkiang, stretches the Takla Makan Desert, an area of 600,000 square kilometers, which also has two subdivisions. The Takla Makan proper, lying in a basin between the Tien Shan Mountains on the north and the Tibetan Plateau on the south, was the site of an Ice Age inland sea the size of the Caspian. North of the Tien Shan, a second, smaller basin holds the Dzungaria Desert. With altitudes typically ranging from 700 to 2,400 meters and much sandy soil and many dunes, it is a desolate, desiccated expanse.

The great Gobi Desert is located northeast of the Takla Makan, partly in China and partly in southern Mongolia. It is a basin and range region, dotted with small mountain ranges and intervening valleys. Much of the ground surface is covered with small stones called *gobi*, the Asian counterpart of the African reg and Australian gibber. The Gobi covers a million and a half square kilometers of what one writer described as a "burning arid waste of dunes interspersed with monotonous rolling expanses of gravel and crossed by occasional ridges of high mountains whose foothills dwindle to low rocky mounds."

From the Ordos Plateau at the eastern end of the Gobi to the Caspian Sea stretches a vast, continuous desert region, interrupted at points by major mountain ranges. Windswept and bitter cold in winter, hot and dry in summer, it has since prehistoric times been a land of nomadic peoples who roam the landscape, tending their sheep and goats and hunting wild animals. The Tatars of northern Turkestan may have been the first to capture and domesticate the wild Przewalski's horse (*Equus caballus przewalskii*) and to invent saddles and trousers. To this day the Mongols hunt foxes with greyhounds and eagles and shoot gazelles during organized hunts in the autumn. This long history of human use has taken its toll on wildlife and habitat in the Gobi and adjacent barrens, as it has in the Thar. The formerly large herds of wild asses, horses, gazelles, saiga antelope (*Saiga tatarica*), and camels have vanished from much of the region or have dwindled to insignificant numbers.

Mobile Life-styles

Most large hoofed mammals of the Asian deserts are nomadic. When rainfall and vegetation are especially abundant over a large area, these animals may spend most of a year in one locale; but in times of meager and widely scattered rains, they may move several hundred kilometers in the course of a year, stopping in one place only as long as food is available there.

The kulan (*Equus hemionus*), or Turkestan wild ass, was once so numerous that it occurred throughout the Kara Kum and Kyzyl Kum deserts, but today it has been reduced to a single herd of 2,000 animals that is usually found in the Badchys Nature Reserve in southern Turkmenistan. The kulan looks almost exactly like the domestic donkey and sounds like the "desert canary" (the feral domestic donkey of North American deserts) when it brays. The Assyrians supposedly tamed wild asses before the horse and used them to draw their chariots.

Like its donkey-gone-wild cousin, the kulan can subsist on coarser vegetation than most hoofed mammals. The kulans of the Badchys were probably highly nomadic when their herds were distributed over the entire Turkestan Desert, and they still migrate in and out of their preserve each year, looking for water but probably also seeking good forage. Like most wild animals, they are more selective about diet than are domesticated beasts. During the breeding season, kulan males become extremely aggressive, biting and kicking ferociously with their hind legs. A victorious male then lays claim to harems of mares and sires most of the herd's offspring.

The Bactrian camel (*Camelus bactrianus*), another animal capable of surviving on very coarse food—including, it is said, leather, blankets, and bones—also moves from area to area. Although its relative, the dromedary (*Camelus dromedarius*), or one-humped Arabian camel, is now wholly domesticated, the Bactrian camel still roams wild in the northern Asian deserts. But it is also widely used as a beast of burden. It is more compact and heavier, with stubbier, stronger legs and longer, heavier hair, than the dromedary. Full-grown at about 17 years of age, it may live 30 to 40 years.

Many desert birds are nomadic. The great Indian bustard (*Choriotis nigriceps*) of the Thar Desert, closely related to the Australian bustard (*Ardeotis australis*), is highly mobile. This bird, which weighs as much as 18 kilograms and stands nearly a meter high, resembles a young ostrich but has a brown back, white neck and belly, and black cap and nape. It does not breed until the summer monsoon, when it migrates to areas that have received rainfall. The cock will then gather a harem of three to five hens and perform its puffed-out, strutting display with drooping wings, while uttering a deep, far-reaching, moaning call. Each hen will eventually lay one or two eggs.

Like the African deserts, Asian deserts have several species of sandgrouse. The imperial sandgrouse in the Turkestan desert and the common Indian sandgrouse (*Pterocles exustus*) in the Thar are well adapted to their environments. Like their cousins in Africa, they are almost entirely seed-eaters and move about for considerable distances from one year to the next, to settle where rain has fallen and new growth of vegetation has produced a seed crop.

The most dramatic spectacle of mass movement associated with rainfall in the desert is that of the desert locust (*Schistocerca gregaria*), a species found across northern Africa, through Asia Minor, and eastward to the Thar Desert. So completely does this animal change its appearance and behavior pattern from one stage of its existence to the next that its two forms were long thought to be distant species. During dry periods this locust is lethargic and solitary, and its color matches the pale desert background. It breeds each year, laying its eggs over the landscape. In dry weather, there is little food for the nymphs that hatch, and few survive; when it rains, the young can feed on abundant vegetation, and many survive. As dry conditions return and vegetation shrinks, the now numerous locusts are crowded together in tiny islands of dwindling vegetation. This crowding stimulates them, so that they become more active and agitated and then pass into the darker, more

Above. *With the Altai Mountains
in the background, a herd of wild
Bactrian camels* (Camelus bactrianus) *crosses the Gobi Desert of
Mongolia. While this species has
been domesticated as a beast of
burden—like its one-humped
cousin, the dromedary camel*
(Camelus dromedarius)—*about
a thousand wild animals still
survive in China and Mongolia,
where they are strictly protected
by both countries. Their humps
are fat deposits that enlarge when
food is abundant and shrink
during periods of drought and
food scarcity.*

Left. *The Asiatic wild ass*
(Equus hemionus), *once abundant
throughout much of Asia, has
been reduced to remnant herds in
western India and the Badchys
Reserve in the Turkmen S.S.R.*

distinct color that marks their gregarious phase. Eventually, the locust swarms increase into the millions, and their agitation reaches such a pitch that, just as a low-pressure weather cell is moving through the region, they launch themselves into the air. Carried aloft by the rising, counterclockwise winds, they move eastward along a storm front. Since low-pressure cells produce rain, the locusts have instinctively launched themselves into a weather pattern that is most likely to provide moisture, plant growth, and food at some point downwind when they descend. Hence they extend their distribution, and some individuals survive and reproduce. The destructive aspect of these mammoth locust swarms is legendary: they devour everything in their path, ravaging cultivated crops, trees, and livestock forage until they eventually dwindle again and resume their solitary phase.

Many animals that live in cold, high-latitude deserts with severe winters combine nomadic movement to areas where rain has fallen with seasonal migration to milder, more southerly climates. The saiga antelope inhabits the semidesert Soviet steppes in summer, migrates in huge numbers each autumn into the northern portions of the Turkestan Desert, then moves north again in spring. The migrations of this homely animal—it has a hump high on its nose and a long, curved snout that ends with two large nostrils set close together, making it look rather like a small elephant with its trunk cut off near the base—are not exactly the same each year. In a drought year, it may move farther north into the less arid plains; in a year of extreme winter weather, it may migrate farther south into the desert.

Before they were wiped out, the wild horses of Central Asia may have followed a similar pattern. They were steppe animals that, wherever located, probably made nomadic movements during each season. In winter they probably migrated south to the desert areas, where lighter snowfall made foraging fairly easy.

An animal tolerates severe winter conditions according to its kind and size. While large hoofed mammals move and forage easily through a few centimeters of snow and can endure subfreezing temperatures, such conditions may tax smaller animals seriously. In winter, many cold-desert birds migrate south to India's Thar Desert, among them quail (*Coturnix coturnix*), imperial sandgrouse, Houbara bustard, short-toed larks (*Calandrella cinerea* and *C. acutirostris*), and desert wheatear (*Oenanthe deserti*). For unknown reasons, still others such as the spotted sandgrouse (*Pterocles senegallus*) and the cream-colored courser (*Cursorius cursor*) fly eastward to the Thar from Africa. (The courser's movements are interesting: it runs rapidly about in zigzag bursts, making sudden stops to look for enemies or to lunge forward for a bit of food.) All these migrants enrich birdlife of the Thar in winter. During their nomadic winter season, many of these visitors gather in flocks and roam the desert in search of areas that have received summer rains, which then produce a crop of edible weed seeds.

Altitude, like latitude, increases the severity of winter. Many denizens of the desert mountain ranges make altitudinal migrations that are similar in purpose to the latitudinal ones. The argali (*Ovis ammon*), a wild sheep of the Central Asian mountain ranges, spends summers in the

Above. *The toad-headed agamid*
(Phrynocephalus maculatus), *a
medium-sized lizard of the
Central Asian deserts, is a
ground-dwelling animal that is
active during the day, when it
feeds on insects.*

Left. *The spiny-tailed lizard*
(Uromastix) *of Iran is unusual
among lizards in its feeding
habits: it consumes plant
material.*

Holson's five-toed jerboas (Alactaga) are solitary, nocturnal animals. As in so many desert mammals, their long ears are heat-radiating organs.

cool comfort of the higher elevations, moves down-slope in fall, lives in the foothills or the desert basins in winter, and then moves back up in spring. In the Gobi region, the Bactrian camel may summer at levels as high as 3,600 meters in the Altai Mountains, but it moves to the desert floor in winter.

Storing for the Future

For small animals that cannot travel great distances to get food, one solution is to gather and store food whenever it is abundant. The Mongolian jird (*Meriones meridianus*), a small rodent of the Gobi Desert, gathers seeds in rainy years, carries them off in cheek pouches, and stores them in chambers in its burrow. These stores, sometimes amounting to several kilograms, keep the animal fed during dry periods.

Seed-eating desert ants also cache food. Ant colonies are made up of several castes. Usually one or a few queens lay the eggs that maintain the colony's population; males of a second caste have the role of keeping the queens bred. A third caste is made up of workers, often themselves subdivided into insects that feed the young and others that forage for food. The foraging workers are the largest single group in the colony, typically numbering in the thousands. Daily they forage over a 10- to 20-meter radius or more around the mound, to bring back seeds for storage. A colony's granary may in time contain a liter or more of seeds.

One bird with unusual caching behavior is the great gray shrike (*Lanius excubitor*), a member of the passerine (songbird) group that is slightly larger than the European blackbird (*Turdus merula*) or the North American robin (*T. migratorius*). Its strong beak, slightly hooked at the tip, reflects its role as a predator that feeds largely on insects, small lizards, and occasionally mice. When it catches its prey, it kills and stores it by impaling it on the thorns of a shrub or tree. In areas inhabited by shrikes, prey such as beetles, locusts, and other small animals can be seen skewered not only on thorns but on the barbs of fences. The black "robber's mask" stripe through the gray shrike's eyes gives it a roguish appearance, which, along with its predatory habits, has earned it the nickname "butcher bird."

One way to store nourishment for a period of hardship is to lay on fat during periods of plenty. The two humps on the back of the Bactrian camel are fat stores, which may weigh up to 10 or 15 kilograms each during times of food abundance. But in periods of scarcity, these large conical growths gradually disappear, until the animal has merely a slightly arched back.

Still another way to avoid starvation during drought or in winter when food is scarce is to slow down the rate at which the body uses energy. In both cold and hot deserts, some rodents become torpid during hot summer droughts. In the cold deserts, they hibernate in winter. A torpid or hibernating animal lowers its body temperature and slows down its metabolic rate to a point where the body is barely functioning. It then uses up stored body fat so slowly that it can live for months in a state of suspended animation without eating.

Among desert animals that hibernate in winter is the long-eared hedgehog (*Hemiechinus auritus*). This quaint, inviting little animal (15 to 18 centimeters long) is an insectivore, related to the moles and shrews. With its pointed snout, short legs, and rounded little body, it looks like a toy. But closer inspection shows that this engaging creature is wrapped in a coat of stiff, sharp spines much like those of porcupines or the Australian echidna. When alarmed, the animal curls up in a ball and lies still. If handled or touched, it straightens its back with a sudden lurch that thrusts its spines into the intruder.

All reptiles in cold deserts are dormant in winter; they burrow into the ground or seek shelter under rocks and logs. The falling ambient temperatures lower their body temperature, and their use of body energy slows accordingly to a glacial pace.

A common means of reducing the risk of starvation when food is scarce in the desert is to diminish or stop breeding altogether. In times of scarcity, the individual that bears many young may overtax its energy reserves and die; an animal that produces fewer young, or none at all, has a better chance of surviving and ultimately producing offspring over a longer span. Desert animals such as Asian gerbils (*Gerbillus*) are among the most flexible breeders of all—reproducing in abundance when there is ample rain but limiting birthrates severely or totally abstaining during drought.

Avoiding Competition

Desert animals have yet another way of reducing the risk of starvation—by avoiding competition between species living together in an area. Two species with identical food requirements can live together as long as their numbers stay at levels that do not exhaust the food supply. If they multiply too greatly, the result may be extinction of one or both species, which is a strong possibility in desert habitats.

Two allied species may coexist if they have diverged to the point where they no longer have the same requirements. A survey of the animal or plant world shows that no two species have exactly the same mode of life and food requirements—that is, in ecologists' terms, occupy the same *niche*. This is probably the result of protracted evolutionary separation minimizing competition.

This pattern can be seen in the animals of the Central Asian deserts. Among the large hoofed mammals of the Kara Kum, the camel generally occupies the lower altitudes and eats coarse food, low in nutrition. The other inhabitant of the lowlands is the goitered gazelle (*Gazella subgutturosa*), a more selective feeder; it nibbles small, annual plants in spring and early summer and then browses on the growing tips of some shrubby species in fall and winter. The wild ass, or kulan, feeds on grasses but also can manage with other low-quality forage. The argali (*Ovis ammon*), an ecological equivalent of the African and Asian ibex, maintains its separation from other species by living in mountainous terrain. The saiga separates itself by living in the steppes during the warm season and entering the northern edge of the desert only in winter.

Predatory species seek different kinds of prey. The red fox (*Vulpes vulpes*) is primarily a rodent-feeder, while the jackal preys primarily on hares (*Lepus tolai*). The wolf (*Canis lupus*), which occurs throughout the northern deserts of Asia, feeds almost exclusively on larger hoofed mammals, especially gazelles. Living in packs of 10 to 25 animals, wolves cooperate in the capture of such larger prey as camels and kulans. Within the pack, individuals are arranged in a hierarchy, with a dominant pair—sometimes termed the "alpha" male and female—topping the order. Breeding activities in the pack are rigidly controlled by the alpha male and female. For periods of several years they may be the only individuals that produce young; during some periods, no offspring are produced.

Birds of prey not only consume different food items, but they also carry on their feeding activities apart from predatory mammals. A big, strong bird, the eagle owl (*Bubo bubo*) feeds on prey as large as hares, bustards, and sandgrouse and does much of its hunting at night, dawn, and dusk. The common buzzard (*Buteo buteo*), active only during the day, feeds on small creatures such as beetles, rodents, or occasionally small birds but is not averse to eating carrion. The harrier eagle (*Circaëtus ferox*), which preys more on reptiles and amphibians, commonly eats frogs, lizards, and small snakes. The black vulture (*Aegypius monachus*) lives on dead animals, sometimes remains left behind by other predators.

The 23 species of lizards living in the Thar Desert clearly demonstrate *resource partitioning*. They differ from one another in dietary preferences and habitat and also are active at different times of day. For example, three species —the Indian sandfish (*Ophiomorus tridactylus*), the large desert monitor (*Varanus griseus*), and the blunt-nosed Indian spiny-tailed lizard (*Uromastix hardwickii*)—prefer similar sandy habitats but have quite different diets. The sandfish feeds entirely on insects; the monitor preys on other lizards (even eating members of its own species), as well as on small rodents and, occasionally, birds; the spiny-tailed lizard is a vegetarian.

A different set of species prefers rocky habitats, and here again the members of this community are separated by their food preferences. The fat-tailed gecko (*Eublepharis macularis*), feeds on insects and sometimes on small lizards. The other big monitor, *Varanus bengalensis*, is a voracious predator that can consume rodents as large as palm squirrels (*Funambulus pennanti*), snakes, and other lizards.

Each species has somewhat different food and habitat requirements, but all require diverse, healthy ecosystems. Many Asian nations are beginning to restore their deserts from their present deteriorated condition. The Soviet Union has set aside 11 desert nature reserves (*zapovedneks*), totaling 625,000 hectares. The Chinese have made great progress in improving the management of desert lands, particularly in stabilizing sand dunes. Perhaps the wildlife of deserts in this part of the world may eventually be restored to greater abundance and become something more than the sparse remnants of another time.

Competition in Cold Deserts of the Americas

The dull crack of massive horns colliding as two desert bighorn rams hurl themselves at each other in early fall . . . the whistled song of the male Western meadowlark in spring . . . the "kerfloop" of the sage grouse cock inflating and deflating the air sacs on its breast. All these sounds are part of rituals that play a vital role not only in mating activities but also in social organization within each animal species of the cold deserts of North America.

Aggressive behavior is a common feature of each social organization. Every species engages in physical combat like the bighorn rams, or else in ritualized fighting like the sage grouse, which may involve more posturing than actual contact, various forms of threat or bluff. This aggressive trait serves to space out the animals over the desert landscape in parcels of land, called *territories* in some species, or leads to a dominance hierarchy ("peck order") in others.

Among some species, territories clearly limit population growth. An overall area of limited size can accommodate only so many separate territories. If through reproduction or migration a population becomes too large, surplus individuals are forced to inhabit the edges, where they constitute an unestablished, nonbreeding excess. In other species, dominance hierarchies appear to promote order. Once established, individuals can go about the task of finding food, though always deferring to dominant individuals.

These behavior patterns are not unique to desert animals. But they do enable arid-land creatures to survive in harsh environments like the cold deserts of the New World.

Cold Deserts of North and South America

The North American cold desert is the New World counterpart of the Old World's Turkestan, Takla Makan, and Gobi deserts. In the western United States it extends from the Canadian–U.S. border on the north to about 37° north latitude on the south, and between the Sierra Nevadas and the Cascade Mountains on the west and the Rocky Mountain chain on the east. Called the Great Basin, this region covers some 400,000 square kilometers.

In these latitudes, and at altitudes ranging from 1,200 to 2,000 meters, temperatures are as severe as those of the northern Asian deserts. The Great Basin is generally dry because it is in the rain shadow of the mountain ranges to the west, as the northern Asian deserts are either situated in rain shadow or simply so far downwind that the westerly winds carry no moisture to them. However, the American cold deserts are not quite so arid as the Asian; their mainly fall-to-spring precipitation ranges from 150 to 300 millimeters per year.

The Great Basin is a region of intermontane basins and scenic mountain ranges. But its vegetation is shrubby, lacking in variety, and not so attractive as that of some other deserts. The dominant shrub in the north is the grayish-green sagebrush (*Artemisia tridentata*), and in the south, shadscale (*Atriplex confertifolia*) is the most common shrub among the lower, sparser salt-desert vegetation. White settlers have been established in the region for little more than 100 years. Though it is grazed by sheep and cattle and planted with wheat or irrigated crops in some areas, the impact of humans has not been so extreme here as that on the Old World deserts. The

110. *Dried winter shadscale* (Atriplex confertifolia) *is scattered over the basins and ranges of the North American Great Basin Desert. Shadscale is one of the many species in the chenopod family that are conspicuous components of cold-desert vegetation in both North America and Asia.*

Above. *Two bighorn rams* (Ovis canadensis) *collide at sunrise in a battle for dominance during the autumnal mating ritual. The successful combatant will dominate the herd in which the animals gather in fall and will mate with most of the females.*

Right. *The desert subspecies of bighorn can be distinguished from the more northerly Rocky Mountain subspecies by its wider, more slender horns, which spiral away from the head. This young ram is probably no more than 5 or 6 years old.*

Overleaf. *The white-tailed jackrabbit (Lepus townsendi) is the only one of the North American jackrabbits that bears a white winter pelage and a brown summer coat. The white phase reflects a northerly, cold-desert distribution and the need for protective coloring.*

most common lowland hoofed mammal is the pronghorn antelope (*Antilocapra americana*), the ecological equivalent of the saiga in Asia. The bighorn sheep (*Ovis canadensis*) is a mountain counterpart of the closely related Asian argali.

The South American counterpart of the Great Basin Desert is the Patagonian Desert, situated largely between 40° and 50° south latitude. Its location to the east of the Andes also makes it a rain-shadow desert in the southern zone of westerly winds. It is farther south than Africa's Cape of Good Hope and thus has no counterpart in Africa. Covering more than 400,000 square kilometers, it is a vast, high, arid plain dissected by numerous east–west canyons with rocky walls, some of which have active streams.

The Patagonian Desert is often swept by cold winds (*los pamperos*), and the southern portion is usually cloudy and foggy. Annual rainfall, coming mostly during the southern winter, is as low as 100 to 125 millimeters in some areas. The land is used mostly for sheep grazing. The most common wild grazing animal is the guanaco (*Lama huanacos*), a relative of the camel and the original wild ancestor of the domesticated llama (*Lama glama*). The rhea (*Rhea americana*) is a close relative of the ostrich, but only about half its height. The puma (*Felis concolor*), known as the mountain lion in North America, occurs throughout Patagonia, where the guanaco is its staple prey.

The Atacama-Sechura

One of the most desolate areas on earth, the Atacama-Sechura Desert is a 3,000-kilometer strip along one-third of the Pacific coast of Chile and all of the Peruvian coast. Located between 32° and 4° south latitude, it is the South American counterpart of the Namib Desert in Africa, the product of exactly the same combination of physical forces. A subtropical high-pressure zone is the basic cause of the aridity at the southern end; the northern portion, lying west of the Andes, is in a rain shadow. Both of these sources of aridity are greatly intensified by cold onshore winds blowing over the icy Humboldt Current. This current flows northward out of the Antarctic along the west coast of South America, just as the Benguela flows northward along the west coast of Africa.

As in the Namib, the result is extreme aridity coupled with extended periods of fog, cloudiness, and moderate temperatures and with humidity that is high for desert regions. Parts of the Atacama-Sechura have the lowest rainfall of any area on earth. The coastal town of Arica near the Chilean-Peruvian border is said to have an average annual precipitation of 1 millimeter—so low that it is all but impossible to measure. Another northern Chilean town, Iquique, is reported to have an annual average of 3 millimeters; yet its year-round relative humidity ranges between 70 and 80 percent.

The Andean cordillera defines the eastern limit of the Atacama-Sechura. The coastal desert strip varies in width, with some portions in Chile divided by north–south coastal mountain ranges that separate narrow marine terraces only a few hundred meters wide from broad inland desert valleys many kilometers across. The entire aspect is desolate: vegetation is sparse or nonexistent, and animal life consists of a miscellany of rodents, foxes, lizards, and insects that somehow manage to survive under these marginal conditions.

To Each His Own
In the New World cold deserts, as in all deserts, food sources are scarce and unpredictable, and like the cold deserts of Asia, winters are severe and pose an additional problem of food supply. Just as unrestrained competition for food between individuals of two species could lead to the extinction of one or both, unrestrained competition between individuals within a single species could cause difficulties as well. Various forms of social organization help prevent this, just as partitioning of resources makes possible the coexistence of two species.

The most visible allocation of resources within a species occurs in the establishment of territories. Individual animals, usually males, occupy and defend tracts of land against the intrusion of other males. A mate is attracted to the territory, breeding takes place, and young are produced within its boundaries. Original determination of boundaries may involve days or weeks of combat between two neighbors; but once established, these limits are sharply defined and well recognized by both combatants. The recognized occupant promptly chases out intruders of the same species.

Territory size varies with the size of the animal. In the North American cold desert, the workers in a harvester ant (*Pogonomyrmex occidentalis*) colony may defend an area of 500 to 1,000 square meters against workers from another colony. Kangaroo rats (*Dipodomys*) in the same desert are solitary, and individuals may defend an area of a hectare or two. Both the ants and desert rats cache food stores, and their territory centers around the burrows holding these caches.

When kangaroo rats square off in combat, they face each other, jump vertically in the air, and lash out with their powerful hind feet. At close quarters, they will also bite with their sharp, chisel-like rodent's teeth. They are so intolerant of each other that, when two male rats are confined together in a small cage and the vanquished cannot escape, the victor will kill him. Males and females join company only at breeding time.

Territory size also varies with the abundance of food. In one North American study, desert iguanid lizards living in areas where their insect food was scarce were observed to claim larger areas; in the process, the lizard population spread out over a wider area.

Predatory animals defend and hunt over much larger areas than do seed or vegetation feeders, because a given area will contain more plant than animal food. The North American badger (*Taxidea taxus*), a member of the weasel family, roams over an area of 2 to 4 square kilometers in the sagebrush desert. Stiff claws and muscular shoulders make this 10-kilogram animal a powerful digger—an ability it uses to tunnel swiftly through the soil in pursuit of burrowing rodents. A single animal can move a huge amount of soil in a short time.

Badgers are nocturnal, hunting and patrolling their territories nightly and spending the days in underground dens. Their territories are a honeycomb of such dens: a single animal may excavate as many as 500 to 1,000 holes in its domain. A badger rarely spends two consecutive days in

The tiny kit fox (Vulpes macrotis), *a delicate, secretive denizen of North American deserts, is largely nocturnal.*

the same den. At dusk an individual will emerge to lumber across the desert during the night, then disappear underground once again at dawn in a den up to 2 kilometers away from the one it used the previous day. Besides the many dens in their territories, badgers commonly dig shallower holes where they cache and partly cover prey to be eaten at a later date.

The more mobile coyote (*Canis latrans*), very similar in appearance, size, and general behavior to the African and Asian jackals, occupies a territory of 40 to 60 square kilometers. The male and female mate in January; the young are born in March. Both male and female hunt and bring food back to their young in an underground den, which is often an enlarged, abandoned badger den. By late summer, the young have grown enough to hunt with the adults, and at night the family group moves about its territory in search of food.

Coyotes are known for their shrill, yapping howl; the American Indians christened them "song dogs." In autumn, a family group may all join in a nighttime chorus that can be heard for several kilometers. By early winter, as both young and adults begin to feel the breeding urge, parents become more aggressive toward and intolerant of the offspring. Eventually, the young are evicted from their parents' territory to disperse in search of mates and unoccupied territories for themselves. Some young may eventually move as far as 50 to 100 kilometers from their place of birth. Meanwhile the adults remain at home and set about housekeeping for a new litter.

The Patagonian puma may range over an area of 100 square kilometers. Like the badger and coyote, it makes nightly forays, moving anywhere from 2 to 10 kilometers at a time. When it kills a large animal such as a guanaco, it eats as much as it can and then covers the remainder with grass, dirt, or leaves, apparently to hide it from foxes and scavenging birds. After making several meals on the remains, the puma resumes its nightly hunting through its territory till it dispatches a new quarry and guards the remnant until it is consumed.

King of the Hill

Establishing dominance, either over territory or within a social group, may involve vigorous combat. In animals other than humans, however, such aggression between members of the same species rarely ends in death. The successful animal is usually larger, stronger, more mature, and more vigorous. Once the stronger animal prevails and the weaker yields, the victor will not continue to attack. Nevertheless, until the contest is settled, the action may be spectacular.

The crunching collisions of two bighorn rams may go on for hours or days. The two rams eye each other, then lunge forward several steps, taking the last one or two while rearing up on their hind legs. Just before colliding, they lower their heads, and the horns absorb the force of the blow. In slow-motion films of two 150-kilogram animals colliding, shock waves are seen to ripple back through the length of each animal's body. The impact can be heard a kilometer or more away. The victor acquires a group of females that breed in fall and bear their lambs in late spring. By this time, the males' mating fervor has subsided; abandoning the company of the females, they

118. *Coyotes* (Canis latrans), *the New World equivalents of the African and Asian jackals, are among the most ubiquitous and adaptable predators in the North American deserts.*
Bottom. *A coyote, as it eyes a coiled rattlesnake* (Crotalus), *is wary of the serpent, for the rattlesnake's bite could be fatal.*

Overleaf. *The badger* (Taxidea taxus) *is a member of the highly variable and adaptable weasel family. An animal weighing about 10 kilograms, it is powerfully built, with a muscular, flattened body and short legs. The front feet are equipped with long, sturdy claws that enable the animal to burrow swiftly through the soil in pursuit of rodents.*

spend the summer in small bands of free and independent cronies with no familial responsibilities. During this time of renewed bachelorhood, there is no hint that these male bighorn rams had been fierce opponents some months earlier.

Pronghorn antelope males fight by lowering their heads, locking horns, and pushing until one subdues the other. As in other fights among hoofed mammals, the contest follows Marquis of Queensberry rules; neither animal takes an unfair advantage of the other or starts thrusting until their horns are properly engaged. Stabbing an animal from behind or driving a sharp horn into the side of an unwary opponent is strictly taboo.

Guanaco fighting is accompanied by loud squealing as each contestant strikes the other with his forefeet and bites the opponent's neck. The winner may gather a group of 100 females.

If a territorial golden eagle (*Aquila chrysaëtos*) invades another's airspace, a dramatic confrontation ensues. The tenant bird seizes the intruder in midair, hundreds of meters above the ground. The two then come wheeling to the ground, locked in each other's clutches, and for some minutes lie there, eyeing each other fiercely. Eventually they separate and fly back into the air, but then may repeat the action. In less spectacular contests, a victorious male rhea acquires a harem of five to six females that will all lay their combined 20 to 40 eggs in a single nest. The male will incubate the eggs and rear the young.

In many cases, combat involves more ritual than actual physical encounter. Two males confront each other, vocalize, strut, and threaten. The smaller, less vigorous individual eventually submits to the larger animal with more bluster. Once dominance is established, the smaller animal "knows its place." Whenever the two meet, a submissive animal automatically yields to the dominant, who reasserts his position with merely a look or a slight head movement.

Probably the most dramatic spring ritual in the Great Basin Desert occurs among sage grouse (*Centrocercus urophasianus*). During a three-month period, each day 20 or more males gather at dawn and dusk on communal dancing grounds, called *leks*. There, each male occupies and defends a territory no more than 10 to 15 meters across. Neighboring cocks pass each morning and evening in elaborate displays: each erects and fans out his pointed gray tail feathers, raises his head and neck, and holds his wings in front of his breast, strutting around with pompous self-importance. He rapidly rubs his wings over the bristle-like feathers that cover his gleaming white breast and makes scratching sounds. He draws his head back over his body and inflates large yellow sacs on his breast. With a sudden forward thrust of his head, he deflates the sacs, expelling the air with a squeaking and slapping sound. The dominant male of the group occupies the central territory of the lek and attracts most of the hens that arrive at the height of the breeding season.

Once the female sage grouse mate, they disperse into the desert to lay their eggs and rear their young. The lek activity wanes by summer, when the males also scatter to enjoy the warmth and abundant food. By late winter or early spring of the following year, the breeding urge rises again, and the leks form once more.

A World of Signals

Defense of territories requires that their limits be marked, and that the aggressive intentions of their tenants be advertised. The individuals of each species have one or more means of communicating this information to its members. Among birds, song is the most common medium. The piping of the Western meadowlark (*Sturnella neglecta*), the high-pitched tinkle of the horned lark (*Eremophila alpestris*), and the hoarse scream of the red-tailed hawk (*Buteo jamaicensis*) circling overhead—all advertise a territory and warn off intruders.

In woody vegetation, territorial occupants will sing while sitting atop a shrub or tree or on a conspicuous limb. Because one function of singing is to advertise a territory and attract a mate, the singer makes himself as visible as possible, fluffing up his feathers and displaying colored or brightly marked patches of plumage. In many deserts, where vegetation is scant and perches for singing are few, birds of many species, especially the larks, sing while circling, hovering, or descending with set wings above their territories.

Birds are acutely sensitive to the songs of other individuals around them; through their sound, they stay in touch with one another. A territorial male meadowlark knows the location of all others of his kind within earshot, which may be 1 to 3 kilometers. He literally lives in a world of song, in voice contact with dozens of other males in his locale. When a territorial male dies or disappears, his place is immediately occupied by another male.

The most active song periods are early morning and late evening; during the middle of the day, the performance slows down or stops completely. At the height of the dawn serenade in the Great Basin Desert, the male mourning dove (*Zenaidura macroura*) will utter his five-note cooing call every 20 seconds. He may keep up this pace for an hour or two after the first pale glint of dawn, but then slows and stops by 10:00 A.M.

In the early stages of the breeding season, when the males are first establishing and defending their territories and attracting a mate, they sing a great deal. Once nesting is well under way, they may stop singing completely in order to avoid attracting predators to the nest. While spring mornings are a din of sound, most late summer mornings are silent.

While mammals also communicate to some degree by sound, they rely heavily on their sense of smell, a sense that birds do not use. Most mammalian species have scent glands somewhere on their bodies—around the face, on the nape of the neck or back, on the legs, or at the base of the tail. Coupled with a delicate sense of smell, ten or a hundred times more sensitive than that of humans, these glands place wild mammals in an environment of scent communication.

The musk gland of the Patagonian hog-nosed skunk (*Conepatus patagonicus*) represents the extreme development of the scent glands found at the base of the tail in other members of the weasel family. Though this skunk uses the gland as protection against enemies, the other weasels (and possibly the skunk as well) use it to mark their territories. Occupants patrol the borders of their territories, rubbing that portion of the body which has the scent glands on rocks, trees, bushes, or bare ground to

The guanaco (Lama huanacos), closely related to Old World camels (but much smaller), is a distant ecological equivalent of the North American pronghorn antelope. The wild ancestors of the domesticated llama (Lama glama), guanacos are the largest grazing animals of the Patagonian Desert.

Vicuñas (Lama vicugna), the second wild member of the camel family in South America, live in the Andes at elevations of 4,000 to 6,000 meters. Standing about a meter high at the shoulder, they have some of the finest fleece found on any mammal. Individual hairs are less than one-half the diameter of the finest sheep's wool.

identify the sites with their odors. Each animal seems to have a unique odor; other members of its species in the area probably recognize it by that odor, just as each bird knows the others of its kind by their song.

Members of the dog family such as the Patagonian fox (*Dusicyon griseus*) and coyote mark their territories by urinating on shrubs and rocks along their boundaries, just as the domesticated dog marks bushes and trees in its own neighborhood. The dog's tendency to roll on a dead animal or other ill-smelling object that it comes upon may be an attempt to mask a foreign odor in its domain with its own scent.

The puma, too, marks its territory with its urine and with its droppings, sometimes covering the latter with dirt. Adult resident males also seem to have the curious habit of marking spots around their territories with what are called *scrapes*. These small piles of soil or dead plant matter, 15 to 50 centimeters long and of equal width, are small signposts placed at the mouths of canyons or on ridges to signal an intruder that this is occupied terrain.

Although many large grazing mammals may not define territorial patterns as clearly as carnivores do, they also scent-mark the areas they occupy. The pronghorn antelope, which has scent glands on its cheeks, marks a shrub by mouthing a branch and then rubbing it with its cheek. The guanaco, a much smaller cousin of the camel, marks its domain with its urine and droppings. It also marks the area with depressions in the ground known as *wallows*. It may spend as much as a half hour in a wallow, pawing it until the soil is loose, rolling on its back with feet in the air, urinating in it, standing up at intervals to neigh, and then rolling about again.

Animals also use visual cues, such as posturing in a variety of ways to threaten a potential trespasser. Lizards use the colored stripes on their sides or underbellies, and birds use colored parts of their plumage, for this kind of signaling. When the animals are threatened, they conspicuously thrust these areas toward the intruder. Many lizards and birds stand as erect as possible, inflating their abdomens or fluffing feathers erect to look as large and formidable as they can.

The Food Pyramid

The abundance of life in an ecosystem can be thought of as being arranged like a pyramid with several layers, each characterized by the source of its organisms' food. Because plants manufacture their own food from sunlight, water, and carbon dioxide, they form the base of the pyramid. Not all plant material gets eaten by herbivores very soon after it is formed; some is coarse, woody material or contains resins and other substances that discourage animals. Not all the vegetation that herbivores eat goes into body growth; some of the energy is used for generating movement, some for maintaining body heat. Generally, in a particular area, there is only about one-tenth as much weight of herbivores as of plant material. The same proportion applies between herbivores and predators. Some plant-eaters die without being eaten, and the predators must also move and maintain body heat; hence the weight of predators will be only about one-tenth that of herbivores.

We can now begin to see why an ecosystem is shaped like

Above. *Laguna Colorado, near the border of Salar de Atacama in Chile, is situated at an altitude of 4,800 meters in the Andes. The lake is salty and red—as its name suggests—from algae.*

Right. *Flamingos (Phoenicopteridae) fly over the surface of Laguna Colorado. These beautiful wading birds are infrequent inhabitants of arid areas.*

a pyramid; the amount of vegetation in an area determines the abundance of the animal layers. In deserts, the small amount of plant growth is the major reason why animal life is so much less abundant there than it is in grasslands and forests. The pyramid effect also explains why grass and shrubs are more prominent parts of ecosystems than are mice and hares, and in turn why mice and hares are more numerous than are foxes and eagles.

Ultimately, the uneaten plant and animal material is consumed by creatures that specialize in feeding on dead matter. Scavenging beetles, termites, tiny mites in the soil, and eventually bacteria play the important role of garbage collectors. Without them, the landscape would be gradually buried under great depths of undecomposed plant and animal material that ecologists call *detritus*. Organisms that eat it are *detritivores*.

Some deserts are so dry that they support almost no permanent vegetation. As we have seen, this is true of parts of the Namib and certainly of the driest areas in the Atacama-Sechura. There, no "normal" pyramid of life can form. The structure that does develop depends heavily on detritus carried in from outside by wind or other means. The predators, rather than relying on a herbivore layer, must build directly on a layer of detritivores. In terms of total numbers, the edifice is extremely meager.

Animal life of the Atacama-Sechura is extremely impoverished; some species become versatile enough to feed on any layer of the pyramid. The opportunist Peruvian fox (*Dusicyon sechurae*) gleans any scraps of food it can find. In winter it may move up into the Andean foothills, where it eats the seeds of shrubs; at this point it is a herbivore. Occasionally, it will catch a rodent and play its traditional role of carnivore. At other times of year, it may move down into the barren coastal desert, where it scavenges dead fish, crabs, or even seaweed washed up on the beaches. Thus it becomes a detritivore and forms a pyramid with but one feeding layer.

The detritus-based animal communities of the coastal deserts are unusual among the world's ecosystems, most of which rest on a base of green plants. The only other major parallels are the communities of abyssal animals in the ocean depths, where no green plants can survive because of lack of sunlight. Like the coastal desert predators, the predator communities of the ocean depths rest on a layer of detritivores that live off dead material formed near the water surface which descends to the depths. Absence of light prevents development of a green-plant base at the bottom of the ocean; in coastal deserts, virtual absence of rain prohibits such a base.

Opposite. *A female trapdoor spider (Ctenizidae) in the Patagonian Desert pulls back the door of its hole at daybreak. The hole of this nocturnal animal, which may be 30 centimeters deep, is lined with the fine silk that it spins.*

Overleaf. *When confronted by an enemy, the Patagonian armadillo (Chaetophractus vellerosus) can curl up within the hard leathery shell on its back to seek protection, like a turtle. The armadillo is closely related to anteaters and sloths.*

Predators and Prey: Hot Deserts of the Americas

The dramas of chase and flight, fang and claw, are very much a part of the desert scene. While the food pyramid decrees that there must be more individual herbivores than predators, more than half of the animal species in the desert are carnivores. Predation is, in fact, one of the factors that keep animal populations within the bounds of their ecosystems. (Social behavior, though also a significant check on population, often is effective only with considerable help from other factors such as predation and severe weather.)

Predation is also a powerful selective force in the evolutionary sense. There is continual pressure on animals to evolve faster means of locomotion, better methods of concealment, or more effective ways of using an area's topography or vegetation for protection. Just as each desert animal is "equipped" with adaptations for avoiding overheating and conserving moisture, so it also has a set of adaptations for eluding predators.

The need to find food is just as powerful an evolutionary force. There is constant pressure on predators to develop more potent hunting skills, such as more acute senses for detecting potential meals and more effective methods of subduing prey. One may see predator-prey evolution allegorically as a never-ending chase in which the prey evolves new means for eluding predators and the predators follow through with new adaptations to neutralize the maneuvers of prey.

The North American Sahara

The North American hot deserts, located entirely in the southwestern United States and northwestern Mexico, are in the Northern Hemisphere subtropical high-pressure zone and therefore constitute the counterpart of the Old World Sahara. But their total area of 900,000 square kilometers is only a tenth that of the Sahara. Although parts of these areas get as little as 25 millimeters of mean annual rainfall, and while the highest air temperature (57°C) ever recorded on Earth was measured in northern Mexico, the conditions in the North American hot deserts are, overall, not so extreme as in the Sahara. Vegetation is more abundant and varied in form; the region has been subject to herding and farming for a much shorter period than Old World deserts have.

Topographically, the region is of the basin-and-range type. So many small-to-large mountain ranges dot these deserts that there are few places without mountains on the horizon. Most of the soils are made up of either fine particles or gravel. Large expanses of sand are infrequent; therefore the stereotype of endless dunes totally without vegetation does not fit most of this region.

The desert region is usually divided into three separate deserts. The northwesternmost and smallest of the three, located in southern Nevada, southeastern California, and western Arizona, is the Mojave Desert. It is the most uniformly austere of the three; its annual precipitation, occurring between fall and spring, measures 100 millimeters or less. Its perennial vegetation, mostly shrubby, has the least structural diversity of any American hot desert; only a few species of yucca and cacti presage the great variety of succulents in the two deserts to the south and east.

The southeasternmost and largest of the three deserts, the

Chihuahuan Desert, extends a short distance into the southwestern United States (Texas and New Mexico) but is situated mostly on the north-central Mexican Plateau between the Sierra Madre Oriental and the Sierra Madre Occidental. With rainfall levels mainly between 100 and 200 millimeters yearly, falling primarily in summer, the region has diverse vegetation, including shrubs, perennial grasses, low trees, cacti, and leaf succulents.

Located between the Mojave and Chihuahuan deserts in the southwestern United States and northwestern Mexico is the Sonoran Desert, generally considered the most attractive desert in the world. Its dual winter and summer rainfall seasons give it the greatest variety of vegetation found in any desert—and perhaps the greatest variety of any habitat, except that of wetter subtropical and tropical regions. Shrubby vegetation is intermixed with cacti in almost endless forms: flat-stemmed prickly pear (*Opuntia*); tubular-stemmed chollas (*Opuntia*), covered with a maze of harsh thorns; waist-high barrel cacti (*Ferocactus acanthodes*), armed with stiff fishhook thorns. Rising out of this 1- to 2-meter stratum of vegetation, lending austere grandeur to the scene, are low trees like the tiny-leafed, green-stemmed paloverdes (*Cercidium*), 3-meter-tall ocotillo (*Fouquieria splendens*) stalks that wear a crown of red blossoms after rains, and towering, 15-meter-tall saguaro (*Cereus giganteus*) and cardon (*Pachycereus pringlei*) cacti. In a year of copious winter rains, a carpeting of brightly colored annuals adds the finishing touch to the beauty of this desert.

The Argentine Monte

The Monte, a small desert region about the size of the Sonoran situated between 27° and 35° south latitude, east of the Andes in Argentina, is South America's counterpart of southern Africa's Kalahari. One would expect the Monte —separated from the North American deserts by the nearly 7,000 kilometers that span much of Central America, the Panamanian isthmus, and the vast equatorial rain forests on the north—to differ even more from the North American deserts than the Kalahari differs from the Sahara. But it is surprisingly similar to the Sonoran Desert, both in appearance and in typical species. Yellow-flowered, green-stemmed paloverdes very like those of the North American deserts punctuate shrubby vegetation dominated by creosote bush (*Larrea cuneifolia*). The dominant North American species, *L. tridentata*, is almost indistinguishable from the second most numerous species in the Monte, *L. divaricata;* in fact, taxonomists formerly thought they were the same species. The Monte has its cardon (*Trichocereus terscheckii*), strongly resembling the Sonoran saguaro, and a variety of other smaller cacti are equivalents of North American species.

There are also similarities in the animals of the two desert regions, the most striking being found among such carnivores as the foxes and cats. Two species of cats—the puma (*Felis concolor*) and jaguarundi (*Felis yagouaroundi*)— occur in both the Sonoran and the Monte. There are, however, differences among the wildlife as well. The complex small-rodent fauna and several species of rabbits and hares of the Sonoran are not duplicated in the Monte; nor are the two species of deer and the peccary, or wild pig (*Dicotyles tajacu*).

Patterns of Flight in Open Terrain

The tactics animals use to escape their enemies and those their pursuers employ are largely determined by vegetation. Dense vegetation prevents animals from escaping or pursuing by means of speed and distance running. In this situation, the pursued are more likely to depend on remaining quiet and slipping off into the vegetation unseen. The pursuers are also more likely to depend on stealth and stalking. On the open terrain of grasslands and deserts, speed is valuable to both the chasers and the chased. In these circumstances, animals have developed the most varied patterns of locomotion.

Among the larger mammals, the Sonoran desert white-tailed deer (*Odocoileus virginianus*) uses the familiar galloping bound, in which both forefeet touch the ground at the same time, followed by both hind feet, the latter actually touching down in front of the former. The white-tail's cousin, the desert mule deer (*Odocoileus hemionus*), flees in great bouncing leaps in which all four feet strike the ground at the same time and then, like springs, propel the animal forward into the next leap.

Desert hares are masters of swift flight. In Mexico, Allen's jackrabbit (*Lepus alleni*), actually a hare and not a rabbit, has perhaps the largest ears of any desert hare. But when the animal is in flight, these heat-radiating organs are drawn back against the body so that they offer no wind resistance or interference with vegetation. While there are no hares in the Monte, there is an ecological equivalent: an overgrown (10 to 15 kilograms) rodent called the Patagonian hare (*Dolichotis patagonum*). This curious animal has a body form and gait like a hare, but its square muzzle and short ears are more like a dog's.

Many desert birds have become partly or wholly earth-bound and rely more on running than on flight. Gambel's quail (*Lophortyx gambelii*) is a swift-footed gamebird of the North American hot deserts that takes flight only when absolutely forced to. In the same deserts, the roadrunner (*Geococcyx californianus*), a member of the worldwide cuckoo family, has abandoned trees for a terrestrial existence. Seldom seen in the air, this predator pursues its prey speedily on foot. Le Conte's thrasher (*Toxostoma lecontei*) is a songbird that prefers scurrying off on foot to flying away from its enemies.

Among desert animals with the most interesting adaptations for movement are those four-legged species which use their hind legs in bipedal locomotion. Several desert lizards have bipedal gaits. The leopard lizard (*Crotaphytus wislizenii*) of North America can rise up on its hind legs and run with great speed, like a miniature of the predatory dinosaur *Tyrannosaurus rex*. When it charges, it must throw its quarry into panic like that which *Tyrannosaurus* inspired in the other dinosaurs it chased bipedally.

A novel form of bipedal gait is hopping on the hind legs, of which the North American kangaroo rats (*Dipodomys*) are classical examples. Like the African and Asian jerboas, these rodents not only bound up to 50 centimeters in one jump on their hind legs, but do so in a zigzag fashion rather like a ball ricocheting between two walls. This curious way of traveling could only be used in the wide-open spaces of a desert. These animals have also evolved a large tuft of black-and-white hair at the end of their long tail, which may be useful as a decoy. As a predator leaps at the

Opposite. The saucy roadrunner (Geococcyx californianus) *is a symbol of desert areas of the U.S. Southwest and northwestern Mexico. A member of the cuckoo family, it spends most of its time on the ground and utters a series of low, cooing notes. This roadrunner has caught a collared lizard* (Crotaphytus collaris).

Overleaf. The collared peccary (Dicotyles tajacu) *is 50 centimeters high at the shoulder and weighs 20 to 25 kilograms. Unlike domestic pigs* (Sus scrofa), *which have four toes on each hind foot, peccaries have only three. They run in packs of 10 to 20 animals. Peccaries have sharp tusks and can be dangerous if cornered or approached too closely.*

conspicuous tail, the animal zigzags so that its pursuer always pounces at the spot the animal just left.

A similar strategy for foiling predators is found among a number of lizard species. The beautiful *Coleonyx variegatus*, the Sonoran Desert representative of the gecko family, has a thick, brightly ringed tail that serves as a fat store. Its bright coloring appears to be a predator decoy, for when a predator grabs the tail, it breaks off quite easily. The predator is left with a mouth full of lizard tail as the *Coleonyx* scampers away tailless but very much alive.

A sandy habitat causes a special set of movement problems. On normal firm surfaces, a snake moves forward by flexing the muscles of its body against the ground. Since its body is bent in a series of arcs, only those portions at an angle to its path of movement create the necessary friction to thrust the body forward. On loose sand, the sand under the arcs merely slips without moving the snake forward. Hence, the sidewinder rattlesnake (*Crotalus cerastes*) has adopted a different movement pattern, which involves throwing an entire loop of its body sideways while setting the full length of its side against the sand and at right angles to its path of movement, thus creating enough friction to move rapidly sideways.

This sidewinding pattern has also evolved independently in the sand viper (*Bitis peringuei*) of the Namib and the sand adder (*Cerastes viperus*) of North Africa. Given a solid substrate, all these species can move forward in the more common pattern of snake movement. The toes of fringe-toed lizards—*Uma notata* of North America and two species of *Liolaemus* in Argentina—facilitate running on sand. This adaptation occurs in lizard species of other deserts, for instance, the fringe-toed sand lizard (*Acanthodactylus scutellatus*) of the Negev and the sand skink (*Scincus philbyi*) of Arabian deserts.

How to Subdue the Quarry

Unlike parasites, which often feed on hosts larger than themselves without killing them, predators dispatch their prey quickly with an arsenal of talons, claws, powerful beaks, and sharp teeth. But despite these weapons, they are generally restricted to feeding on animals their own size or smaller. One way a predator can gain an advantage over larger prey is to develop poisons that dispatch an animal without a physical struggle. Many desert animals have evolved and use this technique.

Scorpions are common in all deserts. With the poison gland and large stinger at the base of a tail curved back over the body, these relatives of the spiders are capable of dealing humans a painful, sometimes fatal sting. Actually, scorpions feed primarily on insects. During the day they seek the shelter of burrows and then emerge at night to lie in wait for hapless prey venturing too close.

Spiders deliver their poison by biting with fangs. They, too, are largely insect feeders, which capture victims in their strategically positioned webs. One of the most frightening-looking desert spiders is the North American tarantula (*Aphonopelma*), a woolly black creature nearly as large as a human hand. This animal can give humans a painful, but not lethal, bite.

Formidable as the tarantula is, it has its own enemies. The large tarantula hawk wasp (*Pepsis*) feeds the spiders to its

Three of the most commonly seen snakes of the North American deserts are not poisonous.
149 top. *The bull snake* (Pituophis melanoleucus), *though very aggressive and pugnacious when first encountered, becomes quite docile after a few minutes of careful handling and stroking.*
Center. *Close relatives of the Western king snake* (Lampropeltis getulus) *that live in the eastern United States are diurnal, but this Western desert subspecies is nocturnal.*
Bottom. *The coachwhip, or "prairie runner"* (Masticophis flagellum), *is a nervous, fast-moving snake that can rarely be tamed. When it bites, nonvenomously, it embeds its teeth and then pulls back, leaving lacerations rather than punctures.*

Overleaf. *The diamondback rattlesnake* (Crotalus atrox) *is one of the few poisonous snakes in the North American hot deserts. The front fangs of this pit viper, which retract and lie flat against the gums when the mouth is closed, raise and point forward when the animal strikes.*

young. Operating exactly like the spider wasp of the Namib, it paralyzes the tarantula with its sting, then drags it to a hole in the ground, where it lays an egg on the victim. When the egg hatches, the larval wasp feeds on the still-living spider. Eventually the spider is killed, but not until the young wasp has fed enough to mature.

The only two venomous lizards in the world are the handsome Gila monster (*Heloderma suspectum*) and its relative, the Mexican beaded lizard (*Heloderma horridum*), two of the larger lizards of North American deserts. Most desert snakes are not venomous; several species subdue their prey by coiling their body around it and suffocating it. Rattlesnakes (*Crotalus*) of North America are quite venomous; their toxin subdues such prey as rodents, which are major parts of their diet, and also contains certain enzymes that, once injected, begin to break down the prey for digestion. Since the snake must swallow its prey whole, this softening action makes consumption easier.

The roadrunner has a similar problem, solved in a less subtle manner. It also eats rodents that are difficult to swallow, especially because of their bones. So, like a butcher pounding a tough piece of meat, the bird may beat a dead mouse vigorously against a rock as many as a hundred times, to break its bones and soften the morsel. Another of the roadrunner's many talents is its consistent ability to kill rattlesnakes. The two animals confront each other, the snake coiled to strike and the roadrunner alert to parry. As the snake lurches forward to strike, the bird hops nimbly to avoid it and then stabs at the snake's head with a lightning thrust of its rapier-like bill. After a few such thrusts, the snake usually lies slain at the victor's feet. Swallowing the head first, the bird may eat a small snake immediately. But if the snake is more than the bird's stomach can accommodate, the roadrunner may swallow only part of it, while running about for some hours with the tail portion hanging out of its bill. Only after digesting the first part, thereby making room for the remainder, does the bird swallow the rest.

As the saguaro cactus has become a plant symbol of North American hot deserts, so this saucy, ubiquitous ground-dwelling cuckoo might be considered their avian symbol. The Mexicans fondly call it *paisano*, or "countryman." Coupled with its jaunty habits is a call unlike that of any other bird. One author says, "Like the bird himself, the sound is derisive, irascible, ribald, threatening, and highly self-confident."

Other carnivorous desert inhabitants are scavengers, meat-eaters that have not entered the evolutionary chase after prey; rather, they eat either animals that have died from other causes or the offal that remains after predators have eaten their fill. The turkey vulture (*Cathartes aura*) epitomizes this group. Its naked head and neck may be an adaptation that allows it to insert its head, unhindered by feathers, into the carcass of a dead animal for some dainty morsel. This bird has developed an odd defense mechanism keyed to its eating habits: if a potential enemy comes too close, the vulture regurgitates its stomach contents on the assailant. Since the vulture's food is usually well putrified when it is eaten, the added rarification it receives in the bird's stomach makes this defense one that potential attackers do not take lightly.

Above. *Velvet mites (Trombidiidae) are colorful members of the North American hot-desert fauna.*

Opposite. *The orange-kneed tarantula* (Brachypelma smithi) *of Mexico is a formidable-looking spider. Its bite, while painful, is not deadly.*

Top row, left. *Unlike many other moths, the amantid moth (Amantidae) is a day-flying animal.*

Center. *The short wings of this common grasshopper (Melanoplus) disclose that it is in a late nymphal stage, not yet a fully developed adult capable of flight. Grasshoppers are characteristically vegetation eaters.*

Right. *This member (Chelinidea vittiger) of the large squash bug family (Coreidae) is a specialized cactus feeder that sucks juice from the plant on which it feeds.*

Bottom row, left. *The thread-waisted wasp (Ammophila) nests in holes that it digs in the ground. Into each hole it places a butterfly larva—which it has stung and paralyzed—and lays an egg on it. The egg hatches, and the larva feeds on the caterpillar.*

Center. *The short wings of this lubber grasshopper (Dactylotum bicolor) do not indicate an immature form; rather, its wings are stunted, for this is a flightless species that can only hop.*

Right. *A blister beetle (Lytta magister) feeds on a cactus blossom. Adult beetles feed on pollen and parts of the flower, but their larvae remain inside the blossom until they can grasp hairs on the legs of visiting bees. The larvae are then transported to bee colonies, where they feed on the bees' eggs, larvae, and pollen.*

Bottom. *A yucca moth (Pronuba) places a ball of pollen on the stigma of a yucca blossom, thus pollinating the plant. The ball is carried in the moth's cupped front legs. After pollination, the moth inserts her eggs in the ovary of the yucca blossom, where they will hatch and the larvae will feed and grow. In this symbiotic relationship, moth and yucca are essential to each other's survival.*

Above. *Audubon's caracara (Caracara cheriway) feeds on a variety of live animals and on carrion.*

But Audubon's caracara (*Caracara cheriway*) is not fazed. This Mexican eagle's handsome looks belie the fact that their owner is a prime candidate for highest honors in undiscriminating dining. This scavenger apparently harasses turkey vultures in flight until they employ their ultimate defense; then the caracara catches the ejecta in midair and makes it his own meal.

Now You See Them . . . Now You Don't
Animals too small or too slow to avoid their predators must somehow elude them by camouflage or by escaping into protective vegetation or terrain. In the sparse vegetation of most deserts, this is not easy. One solution is to burrow underground. The subterranean existence so characteristic of many desert animals solves several problems: as we have observed, in this way these creatures can avoid heat gain, conserve water, and sidestep predators.

But sooner or later, many must come aboveground to feed. Then, like animals that do not burrow, they face the risk of being seen. The desert's bare ground exposes a rich array of colors: black volcanic rocks, red, yellow, and white sandstones, and gray shales. Many desert animals have evolved skin, hair, or feather colors that blend with these backgrounds to give them some degree of concealment. In one area of the Chihuahuan Desert, several species of pocket mice (*Perognathus*) that live in lava beds are black, whereas in nearby areas covered with white gypsum sands the same or closely related species are nearly white. Elsewhere in North American deserts, antelope ground squirrels (*Spermophilus leucurus*) have a beautiful reddish color in locations underlain by red sandstones but a golden hue in sites with yellow sands. Similar matches between skin color and background hues occur with lizards.

Many insects blend visually with the vegetation they feed on, either by matching the color of the foliage or by developing body structures that resemble the leaves or stems. In the Monte, one remarkable animal is a proscopiid insect (*Astroma quadrilobatum*) related to grasshoppers; this stick-shaped creature looks like a stem of the creosote bush on which it feeds.

Succulents Are Their Homes
Myriad species have evolved ways of using plants for shelter from enemies and the elements; for perches on which to rest, display, or sing; and as sources of food and water. Because interiors of large saguaros are soft, birds can easily drill nesting holes in them. The high percentage of predatory species in the desert makes nesting in the open a risky venture. Here, unprotected nests like those typical of birds in temperate forests would seldom (in no more than one in four or five cases) survive long enough to produce free-flying young. Rodents, snakes, or other birds readily see and eat exposed eggs and young. A nest set in a cavity, like those pecked into saguaros, has a much better chance of surviving.

Most holes in saguaros are drilled initially by the lovely gilded flicker (*Colaptes auratus*) and the smaller Gila woodpecker (*Melanerpes uropygialis*). These birds nest in the cavities, but afterward leave their holes available for other birds' use. One of the common beneficiaries of the woodpeckers' work is the charming elf owl (*Micrathene whitneyi*), the smallest owl in the world.

Other bird species choose to build their nests among the stems of the teddy bear (*Opuntia bigelovii*) and jumping cholla (*O. fulgida*) cacti. These plants are covered with a dense maze of spines so incredibly sharp that one cannot even touch the tips gently without becoming impaled; yet the cactus wren (*Campylorhynchus brunneicapillus*), Le Conte's thrasher, and the curve-billed thrasher (*Toxostoma curvirostre*) blithely build their nests in this bristling armory. The latter species actually builds a closed nest of cholla thorns and joints, described by one observer as "ten million cambric needles, set on hundreds of loosely jointed spindles, woven so closely together as apparently to defy penetration of a body however small, but the thrashers go in and out and up through them with the ease of water running through a sieve."

Another species that uses the succulents for building its houses is the white-throated woodrat (*Neotoma albigula*). It piles sticks, scraps of vegetation, small stones, and other miscellaneous items in a globular mass a meter or more in diameter. It often places its nest in the branches of a cholla or between the stiff, spine-tipped leaves of an arborescent Spanish dagger (*Yucca schidigera*). When birds build in chollas, they may make their nests of cholla spines and portions of stem ("joints"). No one knows how these animals and birds that nest in chollas manage to avoid the perilous needles among which they live.

Succulents are also a valuable source of food and water for many species. Peccary hogs (*Dicotyles tajacu*) commonly feed on prickly pear cacti (*Opuntia*), also consuming the spines as they eat the pods. Their digestive tracts must be made of cast iron, for their droppings are a mass of undigested spines. Woodrats also gnaw on prickly pear and yucca leaves, perhaps seeking moisture rather than food. In California, these rats become noticeably more aggressive as summer wears on and vegetation becomes drier and moisture less available. Each rat defends a territory with one or more prickly pear plants and drives away all other small mammals that may intrude.

Many cacti have edible fruit, or "tunas," whose juice is sweet, with a pleasant, piquant taste. The white-winged doves (*Zenaida asiatica*) feed on the saguaro fruit. Hummingbirds, the tiny New World birds that feed on flower nectar, are ecological equivalents of the African and Asian sunbirds and the Australian honey eaters. Costa's hummingbirds (*Calypte costae*) insert their tubular bills into the fruit and drink as they would from the nectar of flowers.

One of the most intimate and novel dependencies which has evolved between a plant and an animal is that of the yucca (*Yucca elata*) and the pronuba or yucca moth (*Pronuba yuccasella*). The timing of the moth's reproduction and the yucca's flowering coincide. When ready to lay its eggs, the moth enters a yucca blossom. There it gathers pollen, which it rolls into a ball, places on the stigma of the flower, and thus pollinates the plant. Over evolutionary time, the flower structure has become so specialized that the plant cannot pollinate itself, nor can it be pollinated by any organism other than this particular moth. Hence the yucca has become completely dependent on the moth for its survival.

But the plant returns the favor. The moth will reproduce only by inserting about six of its eggs into the yucca's

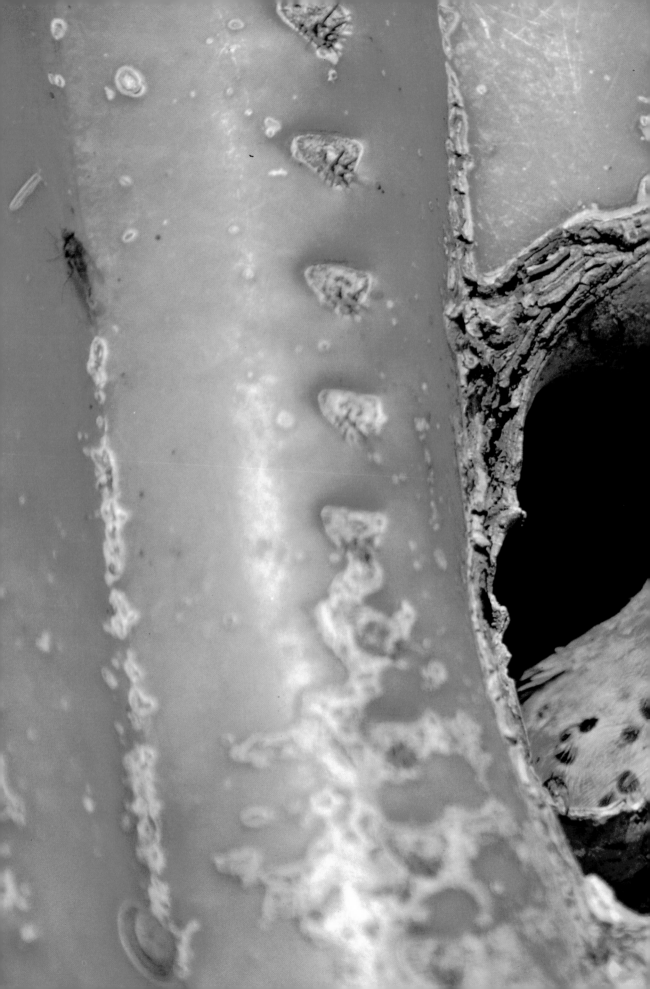

ovary, which develops into the seedpod. After the eggs hatch, the moth larvae live inside the seedpod and feed on the new seeds. By the time the larvae have matured and emerged from the pod, they leave enough seeds intact to allow the plant to reproduce itself. In this way, the plant affords the moth larvae both protection from enemies and food for growth. Since the moth will not lay her eggs in any other plant, the yucca is, in turn, essential to the moth's survival.

They Carry Their Own Protection

Some desert animals have evolved their own protective covering. The porcupine (*Erethizon dorsatum*), normally considered a denizen of the North American forest, is found surprisingly often in the desert. Secure in its own coat of vicious barbed spines, this walking cactus can lumber slowly along the ground, unconcerned about most predators. The porcupine's North American counterpart in the reptile world is the horned lizard (*Phrynosoma*). The numerous species have sharp spines or horns on their heads, over their bodies, and on their tails that make them dubious fare for would-be predators.

While most turtles are aquatic reptiles, many deserts have a land tortoise among their fauna. In North America, the desert tortoise (*Gopherus agassizi*) is a slow-moving animal that eats vegetation and burrows underground for protection from the elements. Like other turtles, it carries a bony shell on its back into which it can withdraw if threatened by a predator. Extremely long-lived for animals, some tortoises survive for 50 to 60 years.

In the Monte, the mammalian counterpart of the tortoise is the armadillo (*Chaetophractus vellerosus*), a species also found in Patagonia. This strange creature, a member of the anteater group of animals that feeds on insects, carries a hard leathery shell over its body much like the bony carapace of the tortoise. When threatened by an enemy, it rolls its body into a ball and covers its head and soft underside with the shell.

This sort of varied, abundant desert wildlife depends not only on a diversity of plant and animal species for food, as in the Central and East Asian deserts, but also on equally varied and plentiful vegetation for shelter. The Gila woodpecker, gilded flicker, and elf owl depend on the saguaro for nesting; without the cactus, these species would disappear. The cactus wren and the curve-billed and Le Conte's thrashers need the cholla cacti for their continued existence; the woodrat and the pronuba moth, the yucca. If we are to continue enjoying the beauties of desert animals, we must protect their diversified desert ecosystems.

Opposite. *The forbidding teddy-bear cactus* (Opuntia bigelovii) *has been described as being armed with "ten million cambric needles, set on hundreds of loosely jointed spindles. . . ."*

Overleaf. *A North American desert tortoise* (Gopherus agassizi) *demonstrates its herbivorous nature. These animals may live to be 50 or 60 years old.*

Converging Evolution: Wildlife of Australian Deserts

Some 120 million years ago, a huge supercontinent called Gondwanaland in the Southern Hemisphere began breaking up into what we now know as India, Africa, South America, Antarctica, New Zealand, New Guinea, and Australia. Except for Antarctica, these continents then drifted toward the continents of the Northern Hemisphere. Australia and New Guinea, last to break away from Antarctica, started their northward and eastward drift a mere 50 million years ago. Hence, for some millions of years, Australia's plants and animals were isolated in what may have been a cold, and surely a moist, latitude comparable to that of Helsinki and Leningrad today. Australian flora and fauna diverged into species very different from those of the rest of the world in adapting to a damp, perhaps even freezing climate.

For the next 40 million years, Australia and New Guinea continued to drift northeastward until they finally came to rest against Southeast Asia. At this juncture, squarely astride the southern zone of subtropical high pressures, Australia experienced blistering heat. The stage was set for an evolutionary test: Would such parched conditions shape the unique Australian wildlife into desert forms? Would their traits converge with those of plants and animals in other deserts?

About 10 million years ago, Australia's northeastward drift had stopped, but water continued to separate the new continent from Southeast Asia. Over a million years ago, at the beginning of the Ice Age, uplifts in the Earth's crust created a chain of Pacific islands that stretches from the tip of the Malay Peninsula at Singapore through such islands as Sumatra, Borneo, Java, and Celebes. The relative closeness of the islands in this chain enabled a ready diffusion of Asian plants and animals. Would the Asian species, used to tropical high temperatures but also to the heavy rainfall and lush verdure of equatorial jungles, be able to adapt to Australian aridity?

Australia's Desert

Much of Australia is reddish in appearance, its soils formed mostly from red sandstones or granites. In the desert, where there is little vegetation to cloak these red hues, reddish ground stretches off in all directions, here and there dotted with plants.

About two-thirds of Australia is arid or semiarid; the total area of its desert is exceeded only by the Sahara. The greatest rainfall occurs at the margins of the continent, whereas the interior third gets 250 millimeters or less of precipitation each year.

Australia's rainfall, more generous than in the driest parts of Africa, Asia, and the Americas, permits a surprising number of small trees to flourish. As nesting, feeding, and perching sites for many species of birds, the trees are important and may in part account for the Australian deserts' considerable variety of bird life.

Except for the Great Dividing Range along the east coast and some low hilly ranges in the center, Australia's land surface features only minor topographic variations. The low hills, occasional rocky ridges, and low cliffs are called *breakaways;* their small caves and crevices provide havens for animals seeking shade—a rich combination of insects, reptiles, and mammals, the largest of which are kangaroos. From the northwestern edge of the continent, a wide

168. *Rainbows like this one over the Simpson Desert in Australia are a rare sight in arid lands.*

170–171. *The frillneck, or frilled dragon* (Chlamydosaurus kingii), *is one of the most spectacular of Australia's lizards. A common inhabitant of woodlands and desert edges, it may rise on its hind legs and dash off at great speed.*

172 top. *This limestone butte in the Painted Desert of Nambung National Park, Australia, has been scored by sand-laden winds for eons.*
Center. *The baked round rocks of some Australian deserts are called "devil's marbles."*
Bottom. *Rounded hummocks of spinifex grass* (Triodia microstachya) *in western Queensland, Australia, which are shaped like devil's marbles, provide shelter for numerous desert species.*

region of sandy soils stretches southeastward into the center.

Other areas have only scant, wispy plant life. Near the center of the continent, a wide gibber plain, called Sturt's Stony Desert, is an arid extreme; only 125 to 150 millimeters of moisture fall here each year. The ground is covered with small, polished stones (gibber), which have a glistening reddish-brown appearance if viewed while one is facing away from the sun, but which appear almost jet black if one is facing into the sun. In summer, as these dark stones absorb and then radiate the energy from a searing sun, the area becomes a natural blast furnace.

Gray-green Leaves of Leather

Long isolation did promote a great deal of evolutionary divergence in Australia's plant and animal life. Whole groups of plants either are unique to Australia or are very different from their kin on other continents. The best examples are the 500 to 600 species of eucalyptus trees spread over the island continent. Although eucalyptus has been introduced and planted all over the world, no other region, except for the Philippines and some islands near Australia, has native species.

Another exceptionally varied group of Australian trees and shrubs whose forms differ from their relatives elsewhere are the acacias. In Australia, there are some 900 or more species of acacia. Unlike other acacias, they are thornless and have simple oblong leaves. Only their flowers and the feathery, compound leaves of the very young plants link them to acacias in other parts of the world.

Despite the evolution of Australian plants into unique groups, they have adjusted to the desert in some of the same ways as plants elsewhere. Many species of eucalypts found, albeit stunted, in most Australian deserts are among the most drought-resistant trees extant. Millions of these have now been planted to provide shade, ornamentals, and windbreaks in parts of North Africa and the Middle East where no other trees will grow. The success of eucalypts and acacias in dry areas is attributable to the leathery texture of their leaves (which reduces transpiration), their huge root systems (which take up moisture from an extensive volume of soil), and the pale color of their leaves (which enables them to reflect sunlight and avoid heat buildup). The pale, grayish-green color of so much of the Australian foliage prompts the common complaint among Australians that their vegetation "is not really green."

In the southern part of the Australian deserts, which receives rain in winter, ground-level shrubs predominate, including saltbush (*Atriplex vesicaria*) and bluebush (*Kochia sedifolia*).

In the northern half of Australia's arid interior, which receives most of its rainfall in summer, the ground-level vegetation is mostly perennial grasses, especially a group of desert grasses called spinifex, another uniquely Australian plant form. Spinifex grasses (*Triodia* and *Plectrachne*) are grayish-green, with very leathery leaves and stiff, sharp-tipped stems. Tufts are 1 or 2 meters across and reach a meter or more in height. These formidable tufts, called *hummocks*, provide secure shelter for an entire community of animals—some predators, some prey. Scorpions, spiders, grasshoppers, desert cockroaches,

175 top. *An adult and two young rabbit-eared bandicoots* (Macrotis lagotis) *feed largely on termites and beetle larvae dug out of the ground with powerful claws. These marsupials sleep while sitting back on their tails, with their heads tucked between their forepaws.*
Bottom. *A small herd of red kangaroos* (Megaleia rufa) *drink at a precious water hole in Sturt National Park, in central Australia. Though most desert animals can survive without free water, they will drink it when available.*

Above. *In the Great Sandy Desert of the Australian interior, as elsewhere, ephemerals or annuals are the first plants to pioneer sand dunes and begin the process of stabilization.*

darkling beetles, elderi geckos (*Diplodactylus elderi*) and hopping mice (*Notomys*) all live together in uneasy truce, apparently preferring each other's company to the greater hazards outside their spiny-tipped fortresses.

Mammals with Pouches

Unique as Australian vegetation is, the continent's mammals have created a greater stir in scientific circles, for they have traveled down very different evolutionary pathways from most other mammals. Yet, when desert animals as unrelated as kangaroos and African gazelles show markedly similar tooth structure, digestive anatomy, and kidney function, scientists have dramatic testimony to the power of convergent evolution. Comparison of kangaroos with Old World and North American hoofed mammals may seem farfetched, but there is no question that—as large, plant-eating animals—all play the same ecological role in their respective deserts.

By 110 million years ago, mammals had diverged into two major divisions. One was the placental mammals, to which most modern mammals belong. Until placental young are well developed, they remain inside the womb for weeks or months. Many species are so precocious at birth that their eyes are open, they are fully covered with hair, and within hours, or even minutes, they are able to move around with their mothers.

The other line of mammals is the marsupials. Their young remain in the womb only a matter of days. When they are born, they are still so poorly formed that they might be considered early embryos. But they are able to crawl feebly through the hair of the mother's abdomen into a pouch on her stomach. There they fasten their mouths to a slender teat from which they get milk. In essence, the bulk of their gestation period is spent in the pouch, so that emergence from the pouch is rather analogous to birth in the placentals.

One Australian desert marsupial, the red kangaroo (*Megaleia rufa*), is about the size of a human being when matured. But unlike a human embryo, a kangaroo embryo stays in the womb for only 33 days. At birth, it weighs less than a gram. Its eyes are undeveloped, and its hind legs are only buds. But its front legs and feet are well developed enough to enable it to claw its way up its mother's abdomen to her pouch, where it instinctively finds a teat. It remains in the pouch approximately eight months and then emerges, fully furred and able to move about with its mother. Its full developmental period is as long as the nine months of a human pregnancy, but the red kangaroo spends only one month inside its mother, followed by eight months in the pouch.

No placental fossils have ever been found in Australia, whereas marsupials appear to have been present on the continent since it was part of Gondwanaland. Consequently, Australia's marsupials evolved very much apart from the placentals of the rest of the world, and no better example of independent evolutionary pathways and divergence can be found. Yet marsupials in Australia and placentals elsewhere have evolved adaptations to similar environments that have resulted in their ending up with amazingly similar appearances, body forms, and internal physiology.

Among the best examples of this convergence are the

The euro, or hill, kangaroo (Macropus robustus) *is a common inhabitant of rocky, hilly areas of the Australian deserts. If it has access to caves or overhanging cliffs, where it can lie in shade and reduce water loss, it does not need fresh water. But if such shelter is scant or inadequate, it will seek water holes periodically during the dry season.*

medium-sized predators. In African and Asian deserts, dog-like predators include wolves, jackals, and hyenas; in North America, they are coyotes and wolves. In Australia, the marsupial equivalent was the Tasmanian wolf (*Thylacinus cynocephalus*), found all over the continent until a few thousand years ago, when early human inhabitants may have almost exterminated it everywhere but Tasmania. By inspecting its underside for the pouch, one learns that this "wolf," so dog-like in its body structure and teeth, is actually a marsupial.

Another marsupial, the euro or hill kangaroo (*Macropus robustus*), prefers the rocky hills and ridges of the breakaways, in this respect filling the same ecological niche as the addax and ibex in Africa and the Middle East, the argali in Soviet Asia, and the desert bighorn in North America. Like them, this kangaroo avoids heat gain by seeking the shade of caves and rocky ledges. Because it depends on this type of terrain, the euro is a sedentary animal, rarely moving far from the rough topography it requires. But since it is sedentary, it is forced to rely on food found near its refuges, which may be very poor during drought. And so the animal has evolved an ability to survive on skimpy foods.

The red kangaroo, on the other hand, is an animal of open terrain, like the Dorcas and rhim gazelles of Africa and the Middle East, the goitered gazelle of Central Asia, and the pronghorn antelope of North America. Like them, it is highly nomadic; it is more gregarious than the solitary euro and usually moves about in small herds. The red kangaroo's penchant for open spaces means that it risks heat gain. Some evaporative cooling occurs when it pants and licks its forelimbs. It replaces the moisture lost in this way by consuming rich, fresh vegetation with a high water content. Its nomadic habits enable it to find such food sources.

There are other parallels between kangaroos and ungulates that live outside Australia. The earliest mammals had 40 to 50 sharp, needle-like teeth to chew their diet of animal matter. In tens of millions of years of independent evolution, the teeth and skull structure of kangaroos and those of the hoofed mammals of the Old and New Worlds have converged into a few, flat-topped teeth for grinding coarse vegetation. Convergence in the stomach structure needed to digest this diet has been equally impressive.

Some other very beautiful Australian desert marsupials are, in one way or another, ecological equivalents of placental mammals in deserts elsewhere. The western hare-wallaby (*Lagorchestes conspicillatus*)—so named by European settlers because of its similarity in size and behavior to Old World hares—is the equivalent of North American jackrabbits and African and Asian desert hares, feeding on plant matter and building nests under shrubs or spinifex tufts. When flushed, this graceful animal bounds out suddenly and dashes away at high speed, just like its Northern Hemisphere placental counterparts.

Unlike true placental mice, the fat-tailed dunnart, or fat-tailed mouse (*Sminthopsis crassicaudata*), is a carnivore that lives primarily on insects. Its name comes from the carrot shape of its tail, a generous store of fat that dwindles to a slender limb during periods of drought and food shortage. In this respect, it is the equivalent of the fat-tailed placental mice of Africa. The Australian wuhl-wuhl

(*Antechinomys spenceri*) is a delicate mouse-sized animal with a pointed nose and a long, tufted tail. It has an engaging way of standing on its hind legs when surveying its surroundings. Using its tail as a rudder to help make abrupt right-angle changes in direction, this animal has become as expert at richochetal locomotion as are North American kangaroo rats and Old World jerboas. One of the oddest mammals in Australian deserts is the rabbit-eared bandicoot (*Macrotis lagotis*). Slightly smaller than a domestic cat, this animal appears to have been assembled from leftover parts. Its long rabbit-like ears are no doubt useful for radiating body heat. The long, fierce claws on its front feet confer a powerful ability to dig, comparable to that of the North American badger. A slender, pointed snout and a banner-like tail, black at its base and white at its tip, combine to give this animal a curious appearance. Bandicoots live in burrows that spiral 1 to 2 meters into the ground. A burrow entrance is often under a spinifex clump, and in the home range of one animal, there may be numerous burrows.

Enter the Rest of the World

After Australia had finished its northward drift toward Southeast Asia and the Ice Age had begun, its fauna ended 50 million years of isolated evolution, and placental mammals arrived from other parts of the world. Over a hundred species of native marsupials had to contend with new competition and predation, as well as with changes in their habitat that these newcomers wrought.

The first wave of ground-dwelling mammals, which came from Southeast Asia during the Ice Age, was made up mainly of rodents related to modern city rats. Since marsupial evolution had apparently produced almost no mouse- and rat-sized plant-eating forms, there was an available niche in Australia's ecosystem for these immigrants. They settled in and developed into some interesting species, with many adaptations familiar in other deserts.

One Australian rodent, the stick-nest rat (*Leporillus conditor*), builds a nest from sticks and rocks. On the ground, the nest is about a meter high. In breakaways, nests are often made in rocky crevices or under overhangs. Here the construction and location of the nest is very similar to those built by the unrelated woodrats (*Neotoma*) in North American deserts.

The spinifex hopping mouse (*Notomys alexis*), though unrelated to the African and Asian jerboas and North American kangaroo rats, resembles them in that it is strongly bipedal and has long ears and a long, tufted tail. It can thrive on a diet of seeds and without fresh water, and excretes the most highly concentrated urine yet found in any rodent.

More recent arrivals, which have had greater impact on the marsupial fauna, were prehistoric humans and their domestic dogs. Aborigines probably first arrived in Australia some 40,000 years ago from Southeast Asia. Over a period of 10,000 to 20,000 years, there were several immigrations, in one or more of which the domesticated dingo dog (*Canis familiaris*) came along. Although aborigines lived a hunter-gleaner existence, they did have spears, boomerangs, and fire; it is fairly clear, then, that after their arrival several marsupials, including a giant kangaroo, a large rhinoceros-like animal (*Diprotodon*), and the

The fat-tailed mouse, or dunnart
(Sminthopsis crassicaudata), *is not
a rodent but a tiny, mouse-sized
marsupial. Its tail is a fat store,
which thickens during periods of
rain and abundant food and
shrinks during drought periods.
Unlike most true mice, fat-tailed
dunnarts of the Australian deserts
are carnivorous, feeding on
insects and other invertebrates.*

Tasmanian wolf, became extinct. Otherwise, prehistoric peoples did not drastically affect wildlife or habitats in this realm.

Their dingos, however, became wild and spread over the entire continent. Now part of Australia's wild fauna, they are predators that feed on kangaroos, rabbits, and other animals. Their sheep-killing habits have enraged ranchers and prompted extensive control programs. With the recent rise of environmental concern, the dingo has become a controversial creature in Australia, reviled by stock-herders but valued by nature lovers.

The most recent immigrations, the nineteenth-century arrival of Europeans and their menagerie of domestic and wild animals, had a more profound effect on Australia's wildlife. Soon Australian deserts witnessed the development of a vast livestock industry, including sheep and cattle, which severely overgrazed the vegetation. A host of domestic animals—horses, camels, goats, pigs, and cats—have gone wild and overrun the landscape, ravaging both vegetation and native animals. European wild animals, especially the red fox (*Vulpes vulpes*) and the Old World rabbit (*Oryctolagus cuniculus*), after having been imported and released, now propagate widely in damagingly large numbers.

These recent assaults have been hard on many marsupials: some are now extinct, others hover on the brink. Still others, such as some of the kangaroos, have grown in number with the opening of the forests and the improvement of pastures for grazing.

Birds, Too, Have Passed the Tests

The evolutionary origins of Australian birds parallel those of the mammals: many species' ancestors go back to the age of Gondwanaland, while many others with uncertain histories have evidently been in Australia also for a very long time. The forerunners of some arrived more recently from Southeast Asia. Because none of these sets of ancestors had evolved under desert conditions, all had to develop adaptations for survival.

Among the primitive forms, the 2-meter-tall emu (*Dromaius novaehollandiae*), second-largest bird in the world, stands out. Its closest relatives are the African ostrich and the South American rhea. Since all are flightless, and thus could not have spread across the oceans, they are one of many pieces of evidence indicating that these three continents were once joined. Like the ostrich and rhea, the male emu incubates the eggs after the female has laid and abandoned them. He seldom leaves them, eats rarely during the eight weeks needed for hatching, and may lose as much as 8 kilograms in the process. Once the young emerge, the male cares for them for as long as 18 months after hatching. He will not mate again during this period; hence most males mate only about every two years.

An even more unusual legacy of Gondwanaland may be the mallee fowl (*Leipoa ocellata*). A member of the general group of birds that includes chickens and pheasants, it is part of a family called "mound builders," found chiefly in Australia and New Guinea. The female lays her eggs in a mound of vegetation and soil, 4 to 5 meters in diameter and 1.5 meters deep, that the male has built. When she comes to the mound to lay an egg, the male pulls back the

vegetation so that she can deposit the egg, and then he covers it. The female lays each egg after a rain, and in a rainy year this procedure may be repeated 30 times.

In the mound, the damp vegetation ferments, producing heat to incubate the eggs. The male is very sensitive to the mound's temperature, which must be maintained at about 33°C. If it rises too high, he may remove vegetation; if it falls, he may add some or cover the mound with soil to hold heat in. After about 49 days the eggs hatch, and the new young must burrow up through a meter of accumulated material. On emerging, they receive no parental care and must make their way alone in the world. Throughout the nesting period, the male remains close to the mound and, fiercely territorial, drives away all intruders.

Australia probably boasts a higher percentage of nomadic birds than any other continent. During summer rains, many emus move toward the center of the Australian landmass and then migrate to southwestern Australia once the fall and winter rains begin. The rare and delicately beautiful princess parrot (*Polytelis alexandrae*) is a desert phantom. A small group of these pastel-colored birds may appear one year and breed, then disappear and not be seen again in that locale for 20 years.

Another pronounced adaptation of Australian desert birds is the way that breeding coincides with rainfall. Following a rain, regardless of the time of year, many species will breed. Small, colorful zebra finches (*Poephila guttata*) will begin nesting within a day or two after a storm. Plumed pigeons (*Geophaps plumifera*), bustards, and the quaint noisy budgerygahs (*Melopsittacus undulatus*) respond almost as quickly.

The Reptiles

Australia's reptile fauna has had a rather different evolutionary history than its mammals and birds. Over half of Australian land mammals are marsupials whose ancestors date back to Gondwanaland days, and less than half are the descendants of placentals that arrived more recently from Southeast Asia. Like the marsupials, a very high percentage of the birds either have ancient origins or else are of unknown origin and may have diverged within Australia long ago during its period of isolation. Only a small percentage have obvious, recent roots in Southeast Asia.

Most of the reptiles, by contrast, derived fairly recently from Southeast Asian relatives. Since ectothermic reptiles are so affected by environmental temperature, they do not flourish in high, cold latitudes. As the Australian "ark" drifted from the Antarctic northward into more tropical climes, it probably did not carry much terrestrial reptile fauna along. But despite a late, single source of immigration, Australia's deserts have the most varied lizard fauna of any. Australian lizards vary in size from small skinks and geckos less than 50 millimeters long to the giant Perentie monitor lizard (*Varanus giganteus*), up to 2 meters long.

These lizards come in many shapes and colors. Most are predators: the smaller ones feed on insects; the larger ones eat other lizards, birds, and small mammals. When threatened, the larger ones are quite aggressive and can bite severely, though not poisonously.

Over time, many lizard species have evolved behavior

184 *top. The brilliantly colored crimson chat* (Ephthianura tricolor), *one of the most nomadic of Australian desert birds, moves great distances to locate an area where rain has fallen. This bird lays its eggs in a cup-shaped nest of grass and twigs, lined with fine rootlets and hair.*

Bottom. *Of the 25 species of doves and pigeons that grace the Australian continent, one-third inhabit arid areas, including this plumed pigeon* (Geophaps plumifera), *which lives on the ground in spinifex. The plumed pigeon has a low metabolic rate, which reduces its need for food and its rate of internal heat buildup.*

186 top. *Amphibians (frogs, toads, salamanders) have soft skins that lose water easily, so they are understandably rare in deserts. The few species present, such as this Holy Cross frog (Notaden bennetti) of central Queensland, are highly specialized in a number of characteristics that enable them to endure drought periods. Accelerated life cycles once water is available and modes of encystment by the adults are but two such adaptations.*
Bottom. *The spiny-tailed gecko (Diplodactylus ciliaris) has rather delicate skin with small granular scales. Most geckos are nocturnal, have no eyelids, and possess curious vertical eye pupils. Their toes are often flattened, webbed, or equipped with disks that facilitate walking on smooth or vertical surfaces or on loose sand.*

Above. *This frog (Cyclorana alboguttatus) of Western Australia shows two adaptations characteristic of arid-land amphibians. It has constructed a tight earthen cell around itself and has secreted a sheath that envelops its body during the dry season. Both coverings prevent water loss.*

189 top. *The armor of the Australian thorny devil* (Moloch
horridus) *reduces its palatability
to many would-be predators. It is
a member of the Agamidae, or
dragon family of lizards.*
Center. *The shingleback, or
stump-tailed lizard* (Trachydosaurus rugosus), *is a member of
the skink family, the largest
family of Australian lizards. The
tail of this denizen of saltbush
deserts is a fat reserve that fuels
its metabolism during hibernation. This shingleback is poised in
an aggressive posture.*
Bottom. *This blue-tongued
lizard* (Tiliqua scincoides) *in
Australia's Simpson Desert is
another member of the highly
diverse skink family, with species
ranging in size from the large
30-centimeter blue-tongue to tiny
legless skinks that slither through
the sand dunes.*
Far right. *An elderi gecko* (Diplodactylus elderi) *peers at the
outside world from the safety of its
spinifex fortress.*

Overleaf. *The large and beautiful
sand goanna, or Gould's monitor*
(Varanus gouldi), *might well
epitomize the reptilian world of
Australian deserts. Goannas
figure prominently in Aboriginal
art, ritual, and mythology. The
"dance of the goannas" is a central
feature of initiation rites in many
indigenous societies.*

patterns and body structures that make them look larger
than they are. In the Australian deserts, the sand goanna
(*Varanus gouldii*) stands on its hind legs when threatening
another goanna, rising up to a height of 25 centimeters or
more. The bearded dragon (*Amphibolurus barbatus*) has a
flap of skin edged with spines just behind the head. When
the animal is threatening, it raises this flap and thrusts it
forward, giving the appearance of a fiercer animal with a
much larger, spiny head. This animal has learned to avoid
extremes of heat. A common midday summer sight in
Australian arid lands is bearded dragons clinging to the
vertical branch of a shrub about 2 meters above the
ground.
One of the most spectacular aggressive displays is that of
the frilled lizard, or frill-neck (*Chlamydosaurus kingii*),
actually a denizen of wooded areas that sometimes
ventures into the desert. It has a wide circular flap of skin
around the neck, folded along the side of the body. During
display, the lizard erects the flap to form a broad collar-like
disk, 12 to 15 centimeters in diameter, around the base of
the head. The animal straightens its legs to stand as tall as
possible and confronts its opponent with frill erected,
mouth wide, and teeth flashing.
Many Australian and North American desert lizards are
ecologically and structurally similar. The Australian thorny
devil (*Moloch horridus*) is both a look-alike and a functional
equivalent of North American "horned toads," or horned
lizards (*Phrynosoma*). Both subsist on a diet of ants.
Another close parallel is that between the Australian
shingleback (*Trachydosaurus rugosus*) and the Mexican
beaded lizard. Both are large, slow, heavy-bodied lizards
with short, weak legs that barely lift their bodies off the
ground as they lumber along.
Lizards of the Australian sandy deserts have adjusted to
living in sand in the same ways as lizards in the Sahara and
the Namib. Many species of burrowing skinks have developed the ability to literally "swim" through sand. Their
legs are tiny, often useless. The legs of one species,
Anomalopus lentiginosus, have become so small that one
must look closely to see them at all. The skinks also have
transparent disks in their eyelids, enabling them to see
when their eyes are closed against the sand.
Australia also has a rich snake fauna, two-thirds of which
belong to the family Elapidae, which includes the cobras of
Africa and Asia and the coral snakes of North America.
Since all elapids are poisonous, bearing their fangs at the
front of their jaws, a very high percentage of Australian
snakes are dangerous. The western brown snake (*Pseudo-
naja nuchalis*), a very aggressive and dangerous elapid
that typically reaches a length of 1.5 meters, is one of the
most common desert snakes. Snakes also engage in aggressive behavior with each other. Like the cobra, the brown
snake flattens its neck to appear larger and more formidable when confronting an opponent.
Of all the world's deserts, those of Australia present some
of the most varied wildlife. This complex ecosystem,
inviting to amateur naturalist and professional ecologist
alike, is also a challenge to conservationists. Let us hope
that the country's vital resources are used wisely, so that
its unique desert habitats may be preserved reasonably
intact.

Picture Credits

Appendixes

Above, 195 top.
*A combination of more solar
energy and faster rotation near
the equator, less energy and
slower rotation toward the poles,
produces six great cells of
atmospheric circulation that ring
the Earth. In one pair of cells, air
rises at the equator, flows toward
the poles aloft, descends at 30°
latitude, and returns to the
equator at the Earth's surface. In
a second pair, air rises at 60°,
flows aloft to 30°, descends there,
and returns to 60° at the surface.
In the third pair, air rises at 60°,
flows aloft to the poles, descends,
and returns to 60° as surface flow.
Zones of descending air at 30°
north and south latitude are called
subtropical high-pressure zones.
Several major deserts are in these
regions of low rainfall.
Surface-flow components of the*

cells have a north-south or south-north orientation. But these are also deflected toward the west because of unequal rotation speeds. The result in the Northern Hemisphere is a northeast-southwest flow between 30° and the equator that creates prevailing easterly winds. Between 30° and 60°, surface winds flow southwest to northeast and become the prevailing westerlies; northeast to southwest flow above 60° produces the polar easterlies. Their mirror image prevails in the Southern Hemisphere.

Prevailing surface winds flowing from oceans to continents carry much of the moisture that falls as rain on the landmasses. Where such onshore winds are forced up and over mountains, they cool and lose their moisture as precipitation. As they descend the down-wind slopes and flow inland, they become warmed. As a result, regions on the lee side of mountain ranges are dry and are said to be in rain shadows. *The Great Basin Desert of North America and Patagonian Desert of South America are rain-shadow deserts.*

195 bottom. *Certain conditions along coastlines can produce almost totally rainless—yet cool and cloudy, or even foggy—conditions. In coastal regions within zones of subtropical high pressure, air descends to the surface and warms; it then moves west out over the ocean. If at the same time a cold ocean current rises to the surface, it will cool a layer of air near the surface and produce a thin layer of clouds or fog (shown as stippled area). Because it is cold, it is heavier than the warm* air above. This weight holds it in place; it is unable to rise and cool further and release its moisture as rain. The result is a cool and cloudy, but almost totally rainless, situation. Such coastal deserts occur along the west coasts of Chile, Peru, Baja California, and the southwest coast of Africa.

The Thermal Environment

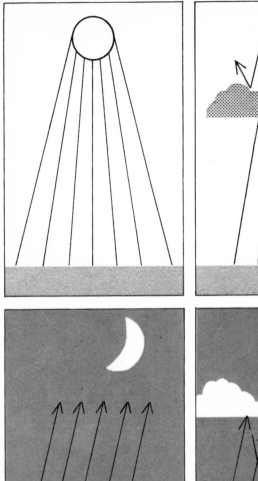

Left. *In desert regions, the lack of clouds and atmospheric moisture allows most of the solar radiation to reach and heat the ground surface. At night this heat is reradiated, and once again the absence of clouds and moisture allows the energy to escape and cool the ground surface. The result is very high temperatures in daytime, and low ones at night. In more humid areas, clouds and moisture intercept much of the incoming radiation, reduce the amount reaching the surface, and prevent daytime extremes. At night the same clouds intercept much of the heat radiating from the ground, reflect some of it back to Earth, and reduce nocturnal cooling. The result is narrower ranges and less extreme temperatures in humid regions.*

197 top. *Burrowers (rodents, ants, etc.) escape midday desert heat by going underground. Temperatures are highest at the ground surface but decline abruptly in the first few centimeters above and below the surface. Desert animals are extremely sensitive even to small fluctuations in temperature, so if a snake climbs into a small shrub, its chances of survival are much greater than if it stays on the ground. The camel has evolved long legs that not only enable navigation of loose sand but also elevate its body to a height where air temperatures are 25 degrees less than on the ground. The many raptors—hawks, eagles, vultures —soar above the desert floor on updrafts of air rising from the surface and reach heights where temperatures are 40–50°C lower. Bottom. Generalized curves represent air temperature, surface sand temperature, and relative humidity in a typical desert environment over a 12-hour period. As the sun climbs higher and the air begins to heat up, the relative humidity, which may have risen to about 50 percent during the night, now drops abruptly to well below 20 percent during the "heat of the day." An incredible increase in the temperature of the sand surface also takes place toward midday, soaring from about 30°C at 8 A.M. to more than 80°C at noon. This daily change in temperature and humidity in desert regions is the most extreme of any natural habitat.*

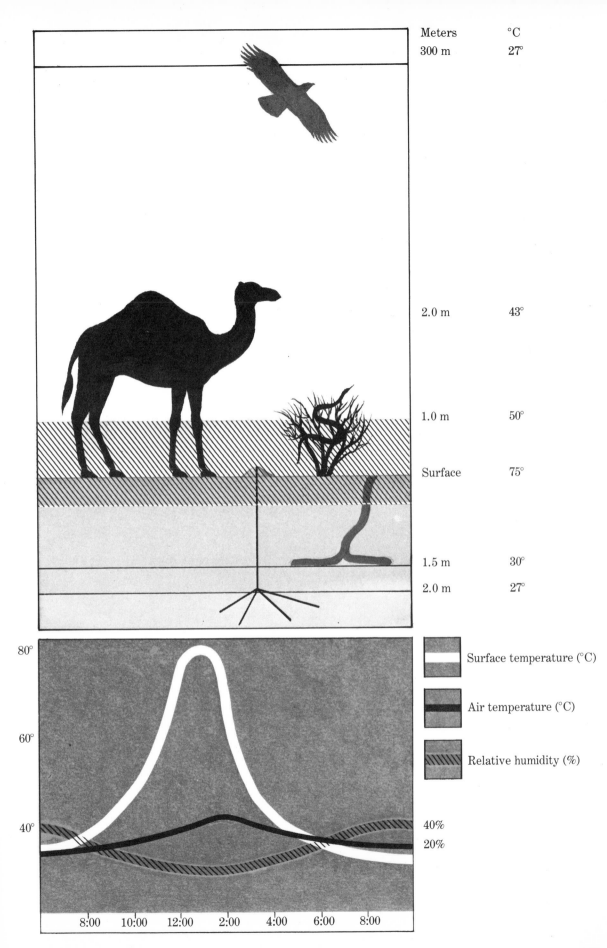

Meters	°C
300 m	27°
2.0 m	43°
1.0 m	50°
Surface	75°
1.5 m	30°
2.0 m	27°

Surface temperature (°C)

Air temperature (°C)

Relative humidity (%)

80°
60°
40°

40%
20%

8:00 10:00 12:00 2:00 4:00 6:00 8:00

Underground Refuges from Heat

In the desert, surface temperatures at midday are usually lethal for any animal, often reaching 60°C (nearly 150°F) or more. Only a few inches underground, temperatures are substantially reduced and much more constant. Moreover, the hot dry air aboveground is very desiccating, whereas the relative humidity below ground is considerably more tolerable. Thus, desert animals such as the brush-toed jerboa (Paradipus ctenodactylus) retreat into burrows during the heat of the day.

Brush-toed jerboa *(Paradipus ctenodactylus)*

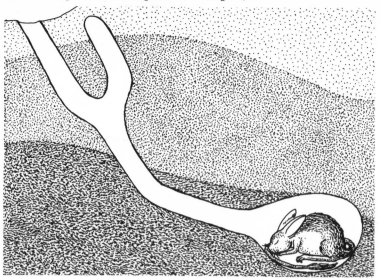

The ecological equivalent of the jerboas in the New World are kangaroo rats (Dipodomys). Both jerboas and kangaroo rats are saltatorial, hopping on their hind legs like kangaroos. Both rodents leave their cool moist burrows only at dusk to feed on seeds at night, when aboveground conditions are less extreme. Jerboas and kangaroo rats have very efficient physiological mechanisms for water conservation (their urine is by far the most concentrated of all mammals). These two desert mammals do not normally drink water but survive solely on metabolic water derived from their foods.

Kangaroo rat *(Dipodomys)*

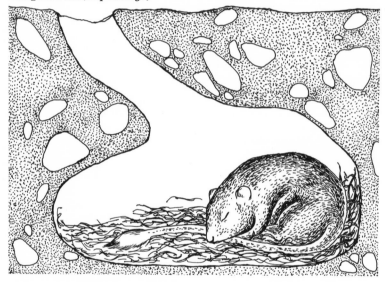

Numerous insects also find refuge below ground. An interesting group is the "honey-pot" ants (Myrmecocystus) of the U.S. Southwest. A special caste known as "repletes" has evolved among these ants. Other workers fill the crops of the repletes with liquid carbohydrates when foods are abundant during moist cool weather. Repletes become so distended they can barely move and simply hang upside-down in underground chambers. During hot, dry weather or other periods of food scarcity, these "living honey casks" serve as reservoirs for the entire colony. One nest in Arizona contained over 14 meters of tunnels and some 1,500 replete ants. Perhaps due to the uncertainty of food supplies, replete ant castes reach their apex in desert regions.

Honey-pot ants *(Myrmecocystus)*

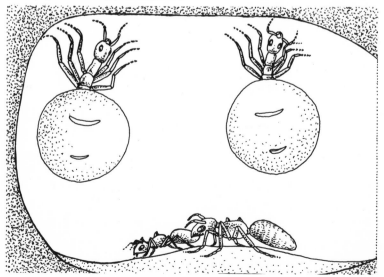

198

Spiders of all sorts abound in most deserts: a head light at night will reveal a glittering array of spider eyes. Many spiders, such as the trapdoor spiders and the Australian goldfield spiders (Ixamatus), dig tunnels. All these spiders use their silk for doors and door hinges, which blend in so well with the terrain that they are virtually invisible unless one sees a spider using them. Like other desert creatures, such spiders use their burrows to escape from harsh desert conditions; many of these burrows are used by other animals as well.

Australian goldfield spider *(Ixamatus)*

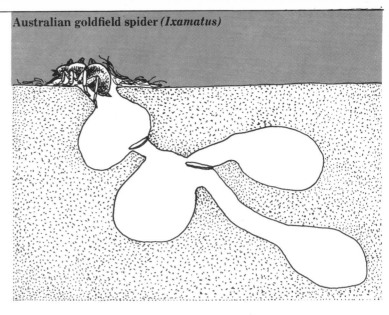

Another burrowing desert animal is the sand roach (Arenivaga investigata) of the American Southwest. This insect does not dig an open tunnel but simply dives and swims through loose sand. Nymphs and adult female sand roaches are wingless and spend virtually all their time below the surface. Adult males, however, are winged; although they also remain below the surface much of the time, males emerge and fly at night. Like many desert insects, these roaches have a remarkable ability to extract water from their environment; they can obtain water from the vapor of unsaturated air. Like all desert creatures also, they minimize water loss by selection of appropriate times and places for their activity.

Sand roach *(Arenivaga investigata)*

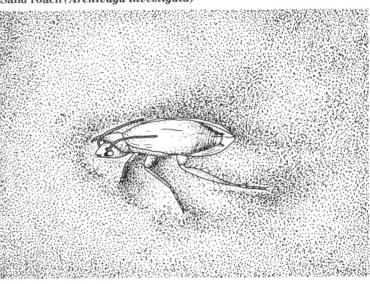

Certain animals spend virtually their entire lives below ground, such as the marsupial golden mole (Notoryctes typhlops) of the Australian sandy deserts. These powerful diggers simply burrow through loose sand in search of buried insects, tubers, and other food. Australian aborigines once left blind members of their bands at Iltoon rockhole in remote Western Australia, at the edge of the Great Victoria Desert, because there was water and an ample supply of marsupial moles, which even a blind person could capture for food. Interestingly, an ecologically almost identical golden mole that has evolved in Africa is a placental mammal rather than a marsupial.

Marsupial golden mole *(Notoryctes typhlops)*

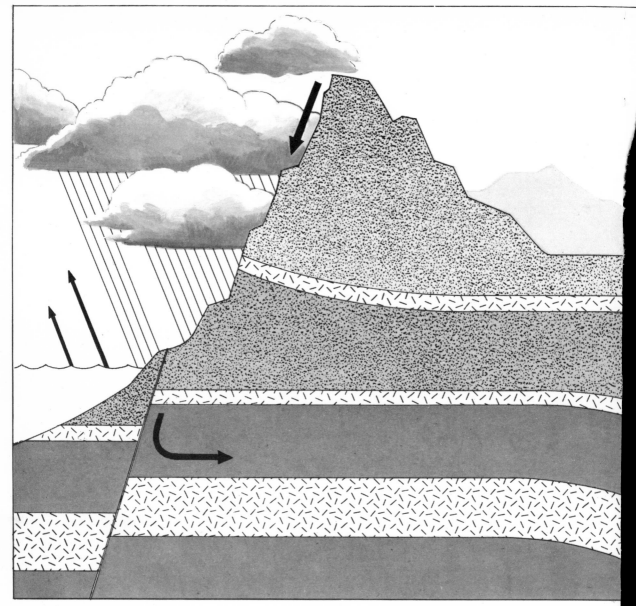

An oasis is an area in the desert that is made fertile by the presence of water. No desert is totally without water; in even the driest region, there is occasional rain. The existence of an oasis is dependent upon a continuous supply of flowing water, which in many instances may travel hundreds of kilometers from its source. As shown in this drawing, the source may be rain that falls on the windward slope of a mountain range. The rainfall percolates into porous rock (water-bearing strata) and travels a great distance, until it is diverted by a fault that blocks its downward movement. The fault itself then becomes a conduit for the fresh water that reaches the surface.

Although the total quantity of water on the Earth's surface does not change, the comparatively small quantity on which life depends is always moving. The "cycle" is a continuous operation. Consider first the rain falling from the clouds onto the slope of the mountain. Much of the rainfall runs off at the ground surface and flows downhill in streams and rivers. Some rainwater is trapped in lakes and ponds; but some water also penetrates the Earth's crust and flows away as groundwater. (And as we have said, it may be this water that continually supplies the oasis.) Eventually all water that has fallen as rain will return to the sea. Water used by plants and animals is returned to the atmosphere through transpiration from the leaves of vegetation and by animal respiration. Evaporation takes place constantly from water surfaces. The water vapor is moved by winds, clouds form, precipitation begins, and the hydrologic cycle continues.

Ephemeral Waters: A Fleeting Habitat

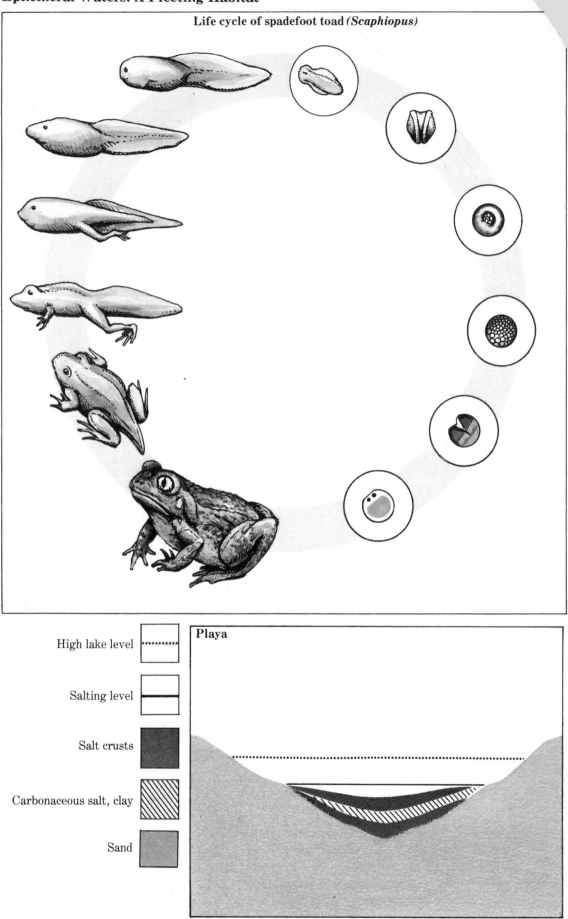

Life cycle of spadefoot toad *(Scaphiopus)*

Playa

High lake level
Salting level
Salt crusts
Carbonaceous salt, clay
Sand

Aquatic organisms would seem to have no place in deserts; but when a rare heavy rain falls in arid lands, dry lake beds fill with precious water and, almost immediately, spring to life. Since such bodies of water soon disappear, their inhabitants must have very accelerated life cycles. Tiny eggs of various crustaceans that have been dormant for months, or even for as long as 25 to 50 years, in the salty crusts of these lake beds (playa) hatch into microscopic larvae. Various species of algae similarly begin to grow and divide, and are eaten by the newly hatched shrimp, which in turn grow rapidly to adulthood, mate, and lay their eggs before the waters evaporate. Such crustaceans as the cosmopolitan brine shrimp (Artemia salina) *either require or are very tolerant of the high salinities and temperatures of many ephemeral bodies of desert water.* Artemia *thrives in waters much saltier than the saltiest seawater. Fairy shrimp* (Eubranchipus vernalis) *occur in fresh water, often in temporary rock pools.* "Tadpole" shrimp (Apus), *also found in ephemeral fresh water, feed primarily on detritus.*

The enormous dry inland lakes of central Australia fill only once or twice in a century, but when they do, ducks and seabirds find their way into the desert, feed and breed on desert shrimp.

Whereas these crustaceans endure as dormant eggs, desert frogs and toads survive droughts as adults encased deep underground in a gelatinous envelope. Using horny projections on its hind feet, the spadefoot toad (Scaphiopus) *digs in backward, secretes its protective covering to reduce water loss, and lies dormant for 8 to 9 months while awaiting late summer rains. When a downpour finally saturates the ground, these toads emerge, mate, and lay eggs in puddles of rainwater. The eggs hatch quickly, and the tadpoles grow and metamorphose into toads within weeks. Unrelated frogs with similar modes of weathering droughts and equally accelerated life cycles evolved in Australian deserts and in the Kalahari Desert of southern Africa.*

Brine shrimp *(Artemia salina)*

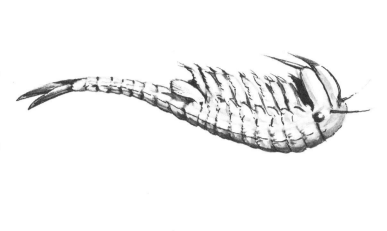

Fairy shrimp *(Eubranchipus vernalis)*

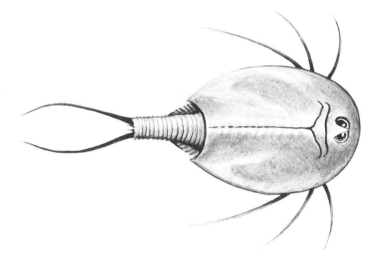

Tadpole shrimp *(Apus)*

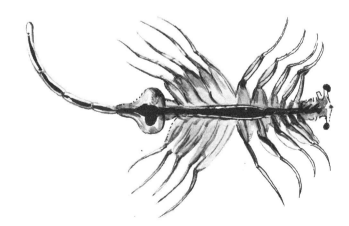

Dunes and the Transporting of Sand

Sand is a powerful agent of erosion. Windblown sand can abrade, polish, and shape the surface of rocks and pebbles, and may wear an unprotected rock down to a virtually flat surface. If a rock is stationary, it will eventually be planed off parallel to the ground, but if its position shifts during the sandblasting or if the wind direction changes, a variety of surfaces can develop.

Sand action

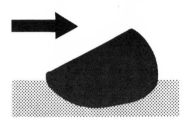

Transported by wind, sand "jumps" along the ground surface; such movement is known as saltation. The impact of windblown, rolling grains may pop other grains into the air. Saltating grains hit the ground surface with force, sometimes knocking grains up to nearly 50 cm from a sandy surface or up to 2 meters on a rocky surface. Sand transported in this fashion builds a variety of wave-shaped deposits—the smaller ones being ripples, and the larger formations, dunes.

Sand movement

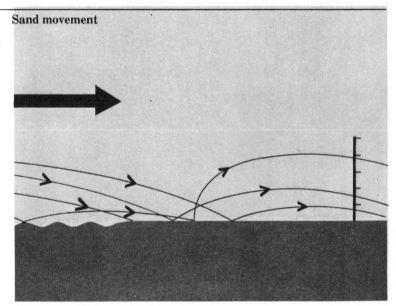

A dune begins to form when an obstacle such as a boulder or a clump of vegetation slows the movement of air and causes an eddy. Windblown sand eventually buries the obstruction, which then catches more sand. Sand is blown up the windward face of the dune and across the top (a). If there is a prevailing wind, dunes characteristically have a gentler windward slope and a steeper lee side. Sand blows up the gentle slope and rolls down the lee side, the so-called slip face (b). Sand accumulates on the lee slope because the wind loses velocity there. When the sand underneath can no longer support this accumulation, it collapses, and a new, lower-angle, stable slope is then formed (c).

Dune formation

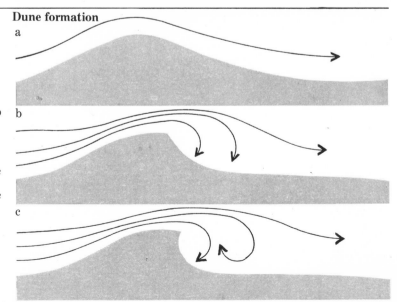

Transverse dunes

Dunes assume a great variety of sizes and shapes. Where sand is abundant and vegetation absent, dunes form in long wavy ridges at right angles to the prevailing wind. Called transverse dunes, these are asymmetric ridges, with gentle upwind slopes and steep downwind slopes (slip faces).

Barchans

Crescent-shaped dunes, or barchans, usually develop according to prevailing wind direction and in areas where sand is not abundant. The crescent shape develops as the prevailing wind blows more sand over a dune's outer edges than across its center.

Star dunes

In areas of shifting wind direction, an isolated dune resembling a star with several points will form. Such dunes have sharp-crested ridges that converge on a central point, sometimes as much as 80 meters high. These "star dunes" are relatively stationary and may remain in place for hundreds of years.

Desert Landforms

The desert is known for its lack of water but is, paradoxically, sculpted by water. Rain, when it does occur, is often heavy, torrential, very intense though lasting but a short time. Desert rains can be so strong that ordinarily dry stream beds quickly overflow with raging currents. Because weathered rock is unprotected by vegetation, it is easily swept away by rushing floodwaters. Tons of sand, pebbles, and sometimes even large boulders churn in the rampaging water that scours deep channels into bedrock and sweeps away everything in its path. There are various distinctive names for these channels. In Latin America and in the southwestern United States, they are called arroyos; in the Sahara and Middle East, they are known as wadis. Typically, an arroyo has nearly vertical walls and a flat floor. As we have said, only at times of rainfall do the arroyos contain water. A canyon, simply stated, is a larger version of an arroyo. Canyons are formed when rivers continue to cut into the land as the land is being uplifted. The result is the creation of the narrow, steep-walled canyons found in many deserts of the world.

Lack of moisture causes desert landscapes to develop slowly. Only the greatest rivers, such as the Nile and the Niger in Africa or the Colorado in the United States, course through the deserts to the sea. Most desert streams are ephemeral; they do not endure for long periods but, rather, dwindle by evaporation or else their waters sink into the ground.

In deserts, drainage patterns are well-developed even though rainfall is slight and runoff rare. Because most desert basins are wholly enclosed by higher land on all sides, drainage is not ultimately connected to the sea. Because of common drainage patterns, deserts throughout the world develop under reasonably uniform weathering processes.

In many deserts, we find great quantities of broken rock debris. All weathered rock materials on the ground will eventually be moved by either the wind, gravity (landslides), or rain wash. In combination, these agents are the most important geological forces sculpting the desert landscape. Wind moves only the fine particles. Landslides occur where relief is great and rain is torrential. The resulting intermittent streams carry sediment from mountaintops to the desert floor.

Exposed rock in the desert is destroyed by mechanical disinte-

Mesa

Arch

Canyon

gration and by chemical decomposition processes. The tremendous daily temperature variation of hot deserts produces forces that expand the rock during the heat of the day and contract it at night when temperatures are cooler. This mechanical process causing the rock to peel apart like an onion skin is known as exfoliation. Though there is little rainfall, there is nevertheless some moisture in the atmosphere —even in the driest deserts. The moisture contacts rock surfaces, and geologists agree that chemical weathering plays an important role in erosion through very gradual dissolution of rock.

Among the most spectacular of desert landforms are mesas. These monumental formations are high structures, capped by resistant rock that overlies the weaker materials which form the steep slopes. Mesas range in size from a few to many square kilometers. A butte is a small version of a mesa, only a few hectares in size. Buttes usually have flat or rounded tops, with steep slopes on all sides. Another fascinating group of desert formations are caprock columns, sometimes also referred to as pillars, needles, pedestals, or pinnacles. Caprock columns are large outcrops that have been worn away more rapidly at the base than at the top, so that the resulting formation has a superstructure that is larger than its supporting base. A vertical system of cracks (joints) usually controls the erosional development of these columns, which vary in shape according to rock type and structure.

Under very special conditions, a hole may be carved through a rock, producing an arch or a natural bridge. Sandstone arches can be cut by rivers. Other arches may form as water enters the cracks and joints of rock and dissolves some cementing material, freeing sand grains, which are then removed by wind and water. As soon as a small hole develops, it can be enlarged by rock falls, rainwater, and wind erosion. A natural bridge can also form when the wind wears away a cliff at the base but not at the top.

Caprock column

Arroyo

Butte

Plant Zonation

1. Century plant *(Agave)*
2. Spanish dagger *(Yucca)*
3. Barrel cactus *(Ferocactus)*
4. Saguaro *(Cereus)*
5. Prickly pear *(Opuntia)*
6. Creosote *(Larrea)*
7. Mesquite *(Prosopis)*
8. Wolfberry *(Lycium)*
9. Cottonwood *(Populus)*
10. Saltbush *(Atriplex)*
11. Pickleweed *(Allenrolfea)*

Nowhere are the effects on plant distribution of soil structure and chemistry, temperature of air and soil, and availability of water more apparent than in the rigorous climate of the desert. Extreme heat and cold fracture the rock of mountains, and torrential rains wash the finer particles down into the lower valleys and basins. Agaves and some yuccas prefer the coarser, well-drained mountain slopes. On the upper, rockier parts of the outwash fans, large cacti such as saguaros (Cereus) *and barrel cacti* (Ferocactus), *may grow, whereas on the finer soil of the lower part of the fans such spreading shrubs as acacia and creosote bush* (Larrea) *commonly dominate. These plants are replaced at the bottom of the fan and on valley floors by mesquite* (Prosopis),

208

6

7

8

9

10

11

wolfberries (Lycium), and sumac
(Rhus), *shrubs and trees that grow
well in silty soil where subsurface
water is available during much of
the year.*
*Where watercourses cut through
the bottomlands, cottonwood
(Populus),* willow (Salix), *desert
willow (Chilopsis), and a recent
immigrant, salt cedar (Tamarix
spp.), find their niches and
provide a wildlife habitat strik-
ingly different from that of the rest
of the desert. Where water collects
and drains away very slowly or
evaporates, the finest soil particles
and salts are deposited.*
*Saltbushes (Atriplex), greasewood
(Sarcobatus), and pickleweed
(Allenrolfea) are among the few
plants that can withstand the
occasional flooding and grow in
the compacted salty soil.*
Although plant zonation in the

*desert is readily apparent, a closer
look reveals that one zone usually
grades into the next. In the center
of the desert, vegetation on slopes
facing north or south might
appear much the same, but along
the northerly edge of the desert, for
example, only the south-facing
slope will have all the characteristic desert plants. (This effect is
reversed in the Southern
Hemisphere.) The cooler north
slope supports plants of the next
major vegetation association to
the north, possibly grasslands or
forests, or those of higher elevations.*
*Throughout the desert, the
higher parts of mountains are
cooler, providing a proper temperature for some desert plants but
not for others. In the still winter
nights, the heavier cold air from
higher areas slips down the slopes*

*and flows through gullies and
canyons, finally collecting in the
lowest spots. Depending on the
extremes of temperature, these low
areas may be inhospitable to cold-
sensitive plants, even though such
low valleys are the hottest places
in the desert during the day.*

Barrel cactus *(Ferocactus acanthodes)*

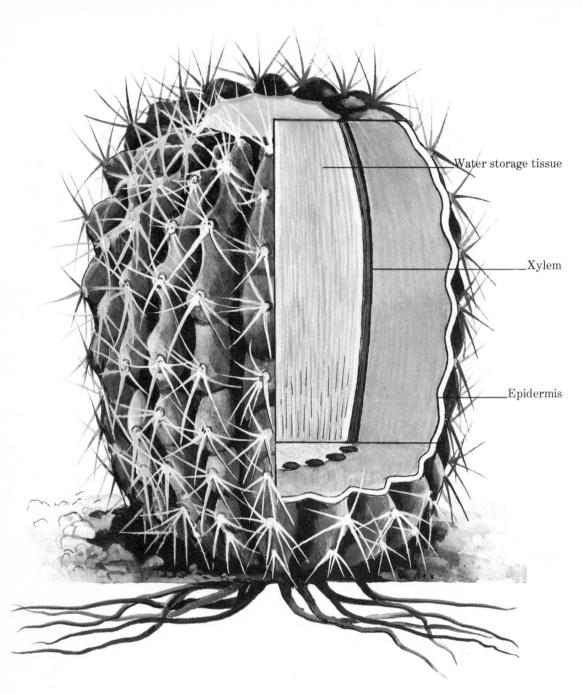

Water storage tissue

Xylem

Epidermis

Left. *The main body of a cactus, whether flat or cylindrical, consists entirely of stem—where, within the green, fleshy interior, production of food by photosynthesis takes place. Food manufactured by photosynthesis is transported to where it is needed or stored by part of the vascular tissue (xylem). Water from the slightest shower is quickly absorbed by shallow extensive roots, transported through the vascular tissue, and stored in the green fleshy tissue. The thick mucilaginous sap of the flesh and thick waxy skin (epidermis and cuticle) covering the entire surface are adaptations for conserving water. Loss of leaves and the transference of photosynthesis to the stem are also water-conserving measures, for the ratio of surface area to volume is reduced. The ridges and nipples present on the surface of many cacti may help to cool this volume while slightly increasing surface, so that more area is struck by light, thereby increasing potential for photosynthesis. The pale hair near the apex of many species serves to keep this sensitive zone of growth cooler by reflecting heat.*

211 top. *Cactus fruits develop from the ovary of the flower and the surrounding fleshy base. Most contain many seeds; in many species these seeds require precisely defined moisture conditions to germinate and establish young plants. In a few species the fruit may also fall to the ground and produce a new plant from sprouts that grow from the fruit wall.*
Center. *Cactus spines attached in clusters at spots along the stems are called areoles. Each areole represents a modified branch that never grows in length. Spines obviously protect the succulent stems from animals that might eat them. They also are believed to help radiate heat from the stem and, especially for those that are pale, to reflect heat from the sun. On species with downward-directed spines, droplets from light rains may accumulate on the spines and drip down near the base of the plant.*
Bottom. *Cactus flowers vary from white to yellow, orange, red, pink, or brilliant red-purple. Those which bloom at night are usually pale and may emit a strong, often sweet odor that attracts nectar-drinking moths in some species and bats in others, which pollinate the flowers during their visits.*

Cane cholla *(Opuntia spinosior)*

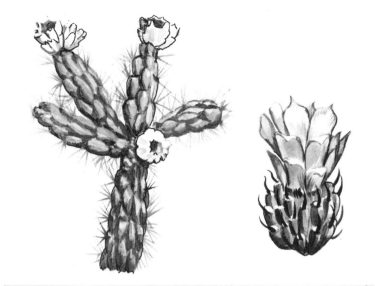

Jumping cholla *(Opuntia fulgida)*

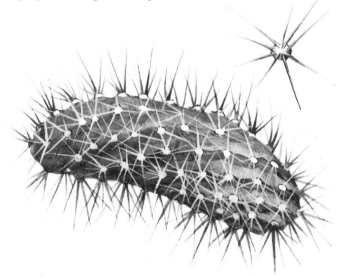

Staghorn cholla *(Opuntia versicolor)*

211

Cacti of Western North America

Several cacti, all in the large genus Echinocereus, are called "rainbow cactus." The name comes from the circles of color the spines produce around the stem, from almost white through gray to several shades of pink, with the bands repeating irregularly. The spectacular flowers, borne on the sides near the top, are seemingly too large for the plant. Colors include bright yellow, yellow-orange, lavender, and pink, or the petals may be banded green at the base, white in the middle, and brilliant pink near the tips. Stems are often single, or sometimes few in a clump, rarely much more than 30 centimeters tall. Particularly gray forms are sometimes called cabeza de viejo, or "old man's head."

Rainbow cactus (*Echinocereus pectinatus*)

Peyote, from the Indian peyotl, has at least seven other Indian names, reflecting its long history in important religious ceremonies. It contains a hallucinogenic alkaloid, mescaline, probably taken from "mescal button," which it is also called (mescal being an Indian word for liquor). "Dry whisky," another common name, refers to its intoxicating effect. This cactus grows in low, broad clumps that barely rise above the ground surface. Individual stems are blue-green and spineless and have only a few low, flat ribs. The distasteful alkaloids, in lieu of spines, protect the succulent flesh from animals that would eat it. Collectors for the drug trade have nearly eliminated the cactus from areas north of the Rio Grande where it once was common.

Peyote (*Lophophora williamsii*)

Pediocactus simpsonii belongs to a small genus restricted to the western United States. Some species are naturally very rare and are now almost extinct in the wild because of collecting. This species, which has a stem up to 13 centimeters in diameter, usually grows singly or a few in a cluster. Unlike most of its brethren, it is able to withstand cold and moisture and is one of the few cacti that may be found at high elevations—up to 3,000 meters. On the other hand, when transplanted to a hot dry climate, it almost always succumbs. The small flowers sit on top, near the center. Color ranges through shades of pink, and some varieties are tinged with yellow.

Simpson's ball cactus (*Pediocactus simpsonii*)

As is characteristic of all members of the genus Opuntia, *the beavertail cactus* (Opuntia basilaris) *has hundreds of minute, barbed spines resembling stiff hair in the areoles, which readily stick in the skin like miniature porcupine quills. It is a relative of other species that produce edible fruit called "tuna" or joints called "nopales," both commonly sold in Mexican markets.*

Beavertail cactus *(Opuntia basilaris)*

Several species in the large genus Mammillaria *have the common name fishhook cactus, given because of the long central spine bent at its tip like a very sharp fishhook. It must be a particularly effective defense, for any animal that might try to eat the stem would be immediately stuck by the straight spines as it nuzzled in, then painfully snagged as it quickly withdrew. Some species are nearly flat, barely visible above the ground surface; others are cylindrical, single or clumped, and most are rather small. Species occur from the western United States to northern South America.*

Fishhook cactus *(Mammillaria microcarpa)*

The scientific name of this Mojave Desert cactus, Echinocactus polycephalus, *roughly translates as "spiny cactus with many heads." The stems branch, but these branches appear as many stems clumped together, as many as 30 in a mound 120 centimeters across. Each stem is usually ball-shaped or cylindrical, rarely more than 30 centimeters tall. The species is found on inhospitably hot, dry rocks and gravelly slopes at low elevations from southern California to western Arizona; it is the northernmost representative of a small North American genus. Its reddish larger spines impart a warm red hue to the plant when viewed from a distance.*

Mojave mound cactus *(Echinocactus polycephalus)*

Succulents in Distantly Related Families

The boojum tree (Idria columnaris), *possibly the most bizarre of plants, has become an image for its native territory, Baja California. Though it looks like something assembled from mismatched spare parts, it has adapted successfully. The tapered trunk stores huge amounts of water, and its scattered branches produce leaves only when ample water is available. It is a close relative of the more widespread ocotillo, or coachwhip (Fouquieria splendens).*

Boojum tree (Idria columnaris)

Among the largest of yuccas, the *name "Joshua tree" was apparently given to this species,* Yucca brevifolia, *by the Mormons who settled in Utah—because of its thick branches, resembling arms upraised in supplication. Like all yuccas, the Joshua tree requires a certain moth for pollination. At night the moth visits its cream flowers, to collect pollen and deposit it on the stigma. It then lays its eggs in the ovary, and the growing larvae eat some of the developing seeds. This is a balanced situation, for some seeds are always left for yucca reproduction.*

Joshua tree (Yucca brevifolia)

The genus Euphorbia *has over 2,000 species, varying from the lowly milkspurges to shrubs and succulents. Succulent species, often perfectly resembling cacti, are native to hot areas of the Old World, especially Africa. They so much resemble a true cactus of the* Cactaceae, *a New World family, that their nature cannot easily be told until they flower. Like the cacti, their fleshy stems provide water storage and minimum surface for maximum volume— a water-saving adaptation. Succulent euphorbias and cacti are remarkable examples of convergent evolution.*

Euphorbia

Nizanda agave is a rather rare
species from the dry rocky slopes
of remote mountains in Oaxaca,
Mexico. It is smaller than most
agaves; other species have leaves
up to nearly 240 centimeters long
and weigh possibly 45 kilograms.
In many, the rosette of leaves
produces a flowering stem only
once, often after many years.
Various species of Agave are used
in producing sisal and henequen
fiber and liquors, and the fleshy
parts may be baked and eaten. The
spines are coated with substances
that produce painful wounds. The
flowers may be pollinated by
nectar-eating bats.

Nizanda agave *(Agave nizandensis)*

The candle plant (Kleinia articulata), as it is called because of its
waxy-smooth cylindrical stems, is
a popular and easily grown indoor
succulent native to South Africa.
It produces rather ordinary leaves
when ample water is available and
loses them during dry times. Like
other fleshy species, it stores water
in stems or in leaves. The latter
often have a small "window" on
one side, to let in light for photosynthesis, whereas much of the
leaf surface is reflective, thereby
reducing heat from the sun.
Kleinia is very closely related to
the huge genus Senecio, the
ragworts, and it is often included
in that genus. Kleinia species grow
in the East Indies, Africa, and
Mexico.

Candle plant *(Kleinia articulata)*

The lily family, to which Haworthia attenuata belongs, is yet
another group that has evolved
succulence as a means of coping
with arid conditions. Haworthias,
all native to southern Africa, are
popular potted plants because of
their neat form and slow growth.
This species has many variants in
the nursery trade. Small whitish-
green flowers are produced on a
slender stalk, which may grow
from the center of the cluster of
leaves. Several species have only a
few leaves, which resemble small
pebbles.

Haworthia *(Haworthia attenuata)*

Desert Lizards: A Diversity of Body Plans

Lerista bipes

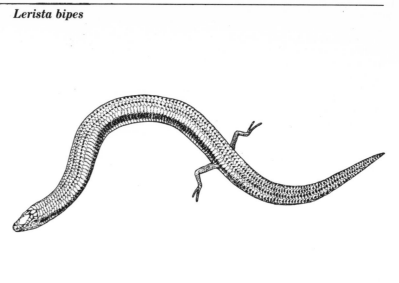

*Lizards are among the most
conspicuous animal inhabitants
of warm deserts. Even Charles
Darwin remarked that they live in
the driest and most inhospitable
places. Mammalogist H. H.
Finlayson commented that the
central Australian deserts are to a
very large extent "a land of
lizards." A key to lizard success in
such hot, arid areas is in their
ectothermy, which allows them to
benefit from a high metabolic rate
when conditions are suitable for
activity, yet permits body temperature to fall when they are
inactive. By retreating into cool
moist burrows, where their
metabolic rates drop during long
harsh periods, lizards conserve
energy and water both daily and
seasonally.
This body heat control enables
desert lizards to fill an amazingly
wide range of ecological niches.
Most species are insectivorous,
although some are herbivorous
and a few eat other vertebrates.
Among insectivores, many eat a
great variety of foods, but some
have specialized diets consisting
solely of ants, termites, or
scorpions. Many species of lizards
hunt actively and widely for their
food, whereas others hunt largely
by ambush, sitting and waiting for
prey. Many lizards are active only
by day, but most geckos and some
skinks are nocturnal.
Some nearly blind, almost legless
skinks lead an underground
existence and eat termites in their
subterranean nests. One such
worm-like fossorial form is*
Lerista bipes *of the Australian
deserts.
Other species are much more like
snakes, particularly the flap-
footed lizards (family Pygopo-
didae) of Australia. Among these
is* Lialis burtoni, *a crepuscular
lizard with keen vision and a large
mouth; found almost throughout
Australia, it eats other lizards
primarily and is ecologically a
"snake." In the same habitats, one
finds the exceedingly tiny, litter-
dwelling skink* Menetia greyi, *a
reptile that is essentially an insect
in its ecology.
Australian desert lizards have
also usurped the ecological roles
filled by predatory mammals
elsewhere: gigantic varanid
lizards, for example, prey on other
lizards as well as on mammals
and nestling birds. One species,
the perentie (*Varanus giganteus*),
is said to reach lengths of up to 240
centimeters. (As recently as 10,000
years ago, a close relative,* V.
prisca, *reached an estimated
length of nearly a thousand centimeters!) Although the smaller
desert pygmy goanna (*V. eremius*)*

Lialis burtoni

Menetia greyi

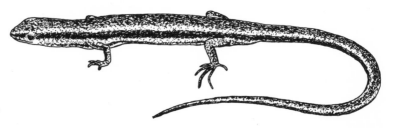

attains a total length of only 40 to 45 centimeters, it too specializes in other lizards for food. Individuals forage over extensive distances each day (up to a kilometer) and, on occasion, subdue other lizards half as large as themselves.

The thorny devil (Moloch horridus) is among the more bizarre creatures of the Australian deserts. These moderately large agamid lizards eat only small ants. Moloch moves slowly and relies on camouflage to escape detection; its sharp spines provide further protection from predators. The function of the spiny appendage on the back of its neck is not known for certain (it is not a fat storage site as formerly speculated), but it may well help to deter predators. When threatened, a Moloch buries its head between its forelegs, leaving this spiny "second head" exposed in roughly the position of the true head. Damaged neck protuberances are seldom seen, but this false head appendage could nevertheless deter predators such as large snakes and varanid lizards, which must swallow prey whole.

A rather different antipredator tactic involving deception is utilized by juvenile Eremias lugubris of the Kalahari Desert. These defenseless small lacertids mimic noxious "oogpister" beetles (the Afrikaans translates euphemistically as "eye squirter"), which emit pungent acids, aldehydes, and other chemicals when disturbed. Adult E. lugubris lizards are pale red, matching the color of Kalahari sand; but the bodies of juveniles are pitch-black with white spotting. Whereas the adults walk with a normal lizard gait, with their backs undulating from side to side, the juveniles walk stiff-legged, with backs arched vertically and reddish tails held flat against the ground. Presumably this makes the tail more difficult to detect. When pursued, young E. lugubris abandon their "beetle walk" and dart rapidly for cover, using normal lizard locomotion. When they reach a size of about 45 to 50 millimeters from snout to vent (nearly the size of the largest oogpister beetles), these lizards "metamorphose" into the adult coloration and permanently abandon the arched walk. The number of broken and regenerated tails is lower in juvenile E. lugubris than among closely related lacertids in the same habitats, suggesting that their beetle mimicry reduces predatory attacks.

Varanus eremius

Moloch horridus

Eremias lugubris

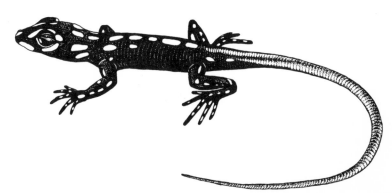

Adaptations to Loose Sand: Fringed Toes and Shovel Noses

An open sandy desert poses severe problems for its inhabitants: (1) windblown sands are always loose and provide little traction; (2) surface temperatures at midday rise to lethal levels; and (3) open sandy areas offer little food or shade or cover for evading predators. Even so, by natural selection over eons of time, lizards have proved able to cope with such deserts. For example, almost all such problems are simply bypassed by subterranean lizards, which avoid both high temperatures and predators by staying underground. Moreover, they actually benefit from loose sand, in that locomotion underground is made easier. Their burrowing has also been facilitated by the evolution of pointed or shovel-shaped heads and countersunk lower jaws, as well as by smaller appendages and muscular bodies and tails. In addition to Australian desert skinks of the genus Lerista (shown in the preceding spread), such worm-like fossorial forms include several species of Typhlosaurus in the Kalahari and Namib deserts, the African genera Acontias and Chalcides, and Ophiomorus in Asian sandy deserts. Such fossorial forms do not occur in the deserts of the New World, although they are found in other sandy habitats, particularly in sand dunes along seashores (Neoseps reynoldsi in Florida and Anniella pulchra in California and Baja California). All but Anniella are in the family Scincidae, members of which have evolved independently in response to the loose-sand environment. Most species feed on termites and bear small litters of large live young.

Similarly, nocturnal lizards, including most geckos as well as some skinks and pygopodids, evade the heat by being active only at night. (They may encounter exactly the opposite problem, since it gets chilly at night in deserts.) Presumably they also find some measure of protection from predators in the darkness. Special adaptations for life at night include elliptical pupils (see the eyes of Lialis burtoni, in the preceding spread, and of Ptenopus kochi and Kaokogecko vanzyli, shown here). Physiological adjustments to cold are also likely. Yet, during the hours shortly after sunrise, but before sand temperatures rise too high, diurnal lizards also scurry about in such habitats. These sand-specialized lizards provide one of the more striking examples of what are called convergent evolution and ecological equivalence. Representatives

Scincus philbyi

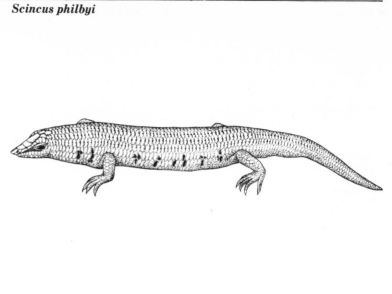

Aporosaura anchietae

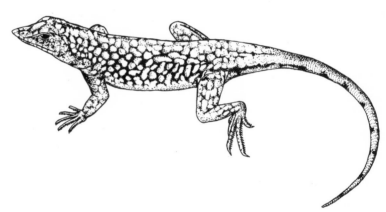

Uma scoparia

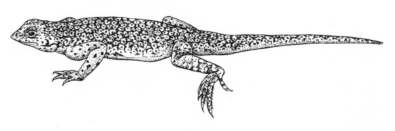

of five different families of lizards scattered throughout the world's deserts have found a similar solution for getting better traction on loose sand; enlarged scales on the toes have evolved independently in skinks, lacertids, iguanids, agamids, and geckos. A skink, appropriately called the "sand fish" (Scincus philbyi), literally "swims" through sandy seas in search of insect food in the Sahara and other deserts, from Algeria to West Pakistan. The Sahara and Arabian deserts also are inhabited by lacertid lizards with fringed toes, such as Acanthodactylus boskianus. *On the windblown dunes of the Namib Desert of southwest Africa, another lacertid,* Aporosaura anchietae, *also feeds on windblown seeds. Several North American fringe-toed lizards, such as the iguanid* Uma scoparia, *usually forage by waiting in the open and eat a fairly diverse diet of insects. They also listen intently for insects buried in the sand, in order to dig them up. They sometimes dash, dig, and paw through a patch of sand and then watch the disturbed area for the appearance of such prey as sand roaches, beetle larvae, and other burrowing arthropods. All three of these species of lizards have flattened, duckbill-like, shovel-nosed snouts, which enable them to make remarkable "dives" into the sand even while running at full speed. They then wriggle along for nearly a half meter or more, while staying fully underneath the sand. Such sand diving is also exploited by the little-known Namib Desert gerrhosaurid* Angolosaurus skoogi. *One must see such a disappearing act to appreciate it fully; it is a most effective means of escape. Most sand lizards bury themselves in sand when inactive during the midday heat and at night—or, in the case of nocturnal species, during the day. Two other species of lizards with fringed toes lack shovel noses: the Australian sandridge agamid* (Amphibolurus clayi) *and the Namib gecko* (Ptenopus kochi).

Some lizards have discovered another way of gaining traction on very loose sands—frog-like webbing between the toes, as is seen in the Palmatogecko rangei *of the Namib and the little-known* Kaokogecko vanzyli *from remote northwestern Namibia, near Angola.*

Amphibolurus clayi

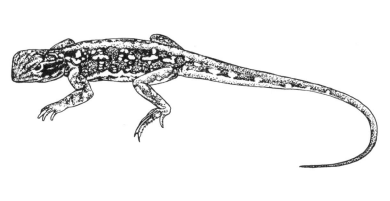

Ptenopus kochi

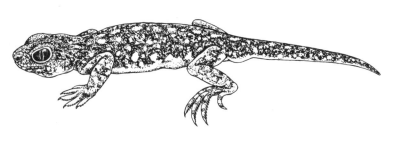

Kaokogecko vanzyli

Tails: Uses and Enigmas

Lizard tails have diversified greatly and serve a wide variety of functions for their possessors. Many climbing species, such as the Australian sand-ridge agamid (Lophognathus longirostris), *have evolved extraordinarily long tails that serve as counterbalances. Tail-removal experiments have shown that such long tails also enable lizards to raise their forelegs up off the ground and run more rapidly on their hind legs alone. Prehensile tails are used as a fifth leg in climbing by other arboreal species such as some geckos* (Diplodactylus elderi) *and the true chameleons* (Chameleo delepis *from the Kalahari Desert).*

Tails of many lizards break off easily. Some species, in a process known as autotomy, can actually lose their tails voluntarily without any external force. Freshly dismembered tails or pieces thereof typically thrash around wildly, presumably attracting a predator's attention while the lizard slips away unnoticed. Certain small predators, such as the pygmy goannas (Varanus gilleni *and* V. caudolineatus), *may actually eat the fragile tails of geckos too large to subdue intact. A skink will return to the site where its tail was lost and swallow the remains of its own tail. Few, if any, other animals display such traits of autoamputation or self-cannibalism.*

Many such lizards possess some special adaptations for tail loss, including weak fracture planes within each tail vertebra, as well as mechanisms for rapidly closing off blood vessels and healing. Losing its tail seems to have little effect on a lizard, and individuals often resume basking and foraging as if nothing had happened. In such lizard species, tails are quickly regenerated. Although regrown tails occasionally are almost indistinguishable from the original externally, their internal support structures are cartilaginous rather than bony. Not all lizard tails are easily broken. Whereas many iguanids have fragile tails, their relatives the agamids do not. Tails of varanids and true chameleons also do not break easily. Lizards with such tough tails cannot regenerate a complete tail if the original is lost. Several members of the Australian gekkonid genus Diplodactylus *(such as* D. ciliaris *and* D. elderi) *have glandular tails that secrete a smelly noxious mucus. When disturbed, these geckos squirt large amounts of this sticky odoriferous material. Surprisingly, the tails of these*

Lophognathus longirostris

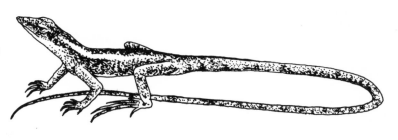

Chameleo delepis

Callisaurus draconoides

geckos are fragile and easily shed, but quickly regenerated.

Tails of some species (especially among juveniles) are brightly colored and very conspicuous, which evidently lures a predator's attack away from the more vulnerable parts of the animal. Thus, when approached by a large animal, the zebra-tailed lizard (Callisaurus draconoides) *of the western deserts of North America curls its tail up over its hindquarters and back, exposing the bold black-and-white pattern underneath, and then wriggles its tail from side to side. If pursued farther, zebra-tailed lizards resort to speed (estimated at up to 20 to 30 kilometers per hour) and long zigzag runs. An Australian desert skink* (Ctenotus calurus) *lashes and shakes its bright azure-blue tail continuously as it forages slowly through the open spaces between plants. Similarly, tiny* Morethia butleri *juveniles twitch their bright red tails as they move around in the litter beneath eucalyptus trees.*

An Australian desert gecko (Diplodactylus conspicillatus) *has a nonglandular but stubby, bony tail. These nocturnal termite specialists hide in abandoned spider holes during the day, and it is thought they point headdownward and use their tails to block off these burrows.*

Another Australian desert lizard with a similar tail tactic is the climbing skink (Egernia depressa). *These lizards wedge themselves into tight crevices in rocks and mulga tree hollows, blocking off the entrance with their strong spiny tails. Spinily-armed tails are also used in this way by numerous other species of lizards, including a Mexican iguanid* (Enyaliosaurus clarki) *and a Saharan agamid* (Uromastix acanthinurus).

Members of another group of Australian lizards (Nephrurus) *possess a unique round knob at the tip of the tail (the desert knobtailed gecko,* Nephrurus vertebralis, *for example). These large nocturnal lizards eat big prey, including other geckos. Both sexes bear the curious knob, but its function remains unknown. Unlike most geckos, their tails are not very fragile.*

Diplodactylus conspicillatus

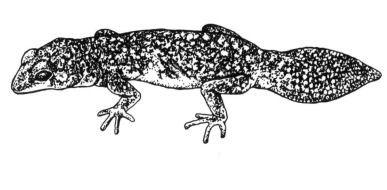

Egernia depressa

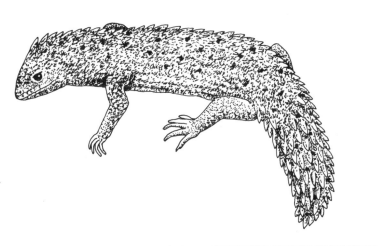

Nephrurus vertebralis

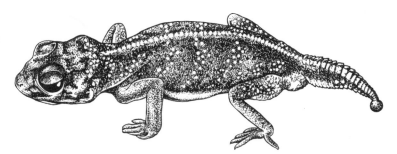

Glossary

Adaptation. An inherited characteristic of a species that is of value to survival in the particular habitat or situation in which the organism normally lives.

Alluvial fan. A large fan-shaped mass of gravel, sand, and silt deposited at the foot of a mountain by a stream that loses velocity as it flows off the slope. Alluvial fans are most visible in deserts, where there is little vegetation to conceal them.

Annual. A plant whose entire life span, from sprouting to flowering and producing seeds, is encompassed in a single growing season. Annuals survive cold or dry seasons as dormant seeds. (See also *Biennial; Perennial.*)

Arroyo. From Spanish. In the southwestern United States and Latin America, a steep-sided, dry stream bed.

Atacama-Sechura Desert. A cold desert located on the Pacific coast of Peru and Chile, characterized by very low annual rainfall, low temperatures, and frequent fog and cloud cover.

Bajada. From Spanish. A more or less continuous band of fused alluvial fans (see) extending along the base of a mountain range.

Barchan. From Arabic. A sand dune formed in the shape of a crescent, whose convex edge faces away from the direction of the prevailing wind; also called a parabolic dune.

Basin and range. A type of desert landscape in which small, parallel mountain ranges alternate with narrow valleys, usually resulting from differential erosion of tilted beds of sedimentary rock, with the mountains composed of harder rock and the valleys made up of softer, more easily eroded rock.

Biennial. A plant whose life span extends to two growing seasons, sprouting in the first growing season and then producing seed and dying in the second. (See also *Annual; Perennial.*)

Butte. A heavily eroded, steep-sided rock formation, generally smaller than a mesa, that is found in arid regions, with a flat or slightly rounded top and composed of horizontal beds of sedimentary rock.

Canyon. A large, narrow, steep-walled valley, generally larger than an arroyo, formed when a stream cuts gradually through layers of sedimentary rock.

Caprock. A layer of hard, erosion-resistant rock overlying one or more layers of softer, more easily eroded rock and usually forming the top of a mesa or butte.

Chihuahuan Desert. A hot desert located in southern Texas and New Mexico and on the northern part of the Mexican Plateau, characterized by having between 100 and 200 millimeters of rainfall annually, falling mainly during the summer months, and having a moderate variety of plants.

Convergent evolution. Evolutionary change in which two or more unrelated organisms come to resemble one another as a result of acquiring similar adaptations to similar conditions.

Detritivore. An animal feeding largely or entirely on detritus.

Detritus. Debris consisting of small bits of dried plant or animal matter, often occurring in deserts in wind-deposited accumulations. Detritus is an important food source in deserts, where decay is slow and where there are few other sources of food.

Dune. A hill or mound of loose, windblown sand.

Dzungaria Desert. A cold desert occupying a vast basin north of the Tien Shan Mountains in the western Chinese province of Sinkiang; usually considered a subdivision of the Takla Makan Desert (see).

Ecological equivalents. Two or more animals that occupy similar ecological niches in different geographical areas. Examples are the sidewinder rattlesnake of American hot deserts and the sand adder of the Sahara, both of which have the same sidewinding mode of locomotion on loose sand.

Ecology. The scientific study of the relationships between plants, animals, and their nonliving environment.

Ecosystem. The total pattern of relationships between plants, animals, and their nonliving environment in a particular area or habitat. The particular position of any species in this pattern is termed its niche.

Ectotherm. An animal whose body temperature is dependent upon that of its environment and is not maintained at a constant

level by internal metabolism. (See also *Endotherm.*)

Endotherm. An animal whose body temperature is maintained at a constant level by internal metabolism and is more or less independent of the temperature of the environment. Of all animals, only birds and mammals are endotherms. (See also *Ectotherm.*)

Erg. From Arabic. Any of several very large sandy tracts in the Sahara Desert, of which the Great Sand Sea in Egypt and Libya—the size of France—is the largest.

Erosion. Gradual wearing away of rock or any other surface feature by the abrasive action of water or wind.

Evolution. The change of organisms through geological time as a result of mutation and the gradual acquisition or perfection of adaptations.

Food pyramid. A graphic representation of the food-relationships among organisms in an ecosystem, in which the organisms that form the primary source of food have the largest population and are considered the base of the pyramid, and in which the population decreases with each level of consumers until one reaches the highest level of predators, which have the smallest population and are considered the apex of the pyramid.

Gibber. In Australia, a desert area whose surface is covered with polished stones and boulders. (See also *Reg.*)

Gobi Desert. A large cold desert in northern China

and southern Mongolia.

Gondwanaland. A very large landmass that formerly existed in the Southern Hemisphere, which, according to modern concepts of continental drift, began to break up about 120 million years ago and whose fragments form the present-day continents of South America, Africa, Australia, and Antarctica and the subcontinent of India.

Hamada. From Arabic. In North Africa and Arabia, a desert area whose surface is covered with flat pieces of calcareous rock or limestone. (See also *Reg.*)

Herbivore. An animal that feeds primarily on plants.

Iranian Desert. A large cold desert located in Iran, Afghanistan, and western Pakistan.

Kalahari Desert. A hot desert in South Africa, Botswana, and Namibia, characterized by sparse rainfall concentrated in the summer months and by large areas of parallel sand ridges separated by shallow depressions that catch and hold rainwater for brief periods.

Kara Kum. The southwestern portion of the Turkestan Desert (see), located in the southern Soviet Union east of the Caspian Sea and south of the Aral Sea and characterized by large areas of predominantly blackish sand.

Kyzyl Kum. The northeastern portion of the Turkestan Desert (see), located in the southern Soviet Union, south and east of the Aral Sea, and characterized by large areas of predominantly reddish sand.

Mesa. A steep-sided, more or less flat-topped rock formation, generally larger than a butte, composed of horizontal layers of sedimentary rock and found in arid regions.

Mojave Desert. The driest, hottest, and most barren of the hot deserts of North America, located in southern Nevada, southeastern California, and southern Arizona and having less than 100 millimeters of rainfall annually.

Monte. A hot desert located along the eastern edge of the Andes in northwestern Argentina, notable for its floral and faunal similarity to the Sonoran Desert (see) of the southwestern United States.

Namib Desert. A cold desert located in a narrow strip along the Atlantic Coast of Namibia, formerly South-West Africa, characterized by very low annual rainfall, low temperatures, and frequent fog or cloud cover.

Negev Desert. A small hot desert located in Israel that is an extension of the hot desert system of North Africa and Arabia.

Niche. The particular position of an organism in its ecosystem (see), with reference to the ways in which it affects or is affected by other organisms and its nonliving environment.

Nomadism. The habit, in some animals, of performing large-scale irregular movements, often in herds or flocks, usually in search of shifting sources of food or water.

Oasis. An isolated moist or wet area in an otherwise arid desert region.

Oed. From Arabic. In North Africa and Arabia, a dry gully or stream bed; another form of the Arabic word *wadi*. (See also *Arroyo*.)

Patagonian Desert. A cold, rain-shadow desert located east of the Andes in southern Argentina.

Perennial. A plant whose life span extends over several growing seasons and that produces seeds in several growing seasons, rather than only one. (See also *Annual; Biennial*.)

Playa. From Spanish. A dry, or usually dry, lake bed in a desert, often containing deposits of salts left behind by evaporation.

Qas. From Arabic. In North Africa and Arabia, a dry lake bed. (See also *Playa*.)

Rain shadow. An arid area on the leeward side of a mountain range, receiving little rainfall because the prevailing winds have lost their moisture in passing over the mountain range.

Reg. From Arabic. In North Africa and Arabia, a desert area whose surface is covered by polished stones and boulders. (See also *Gibber*.)

Rub' al Khali. Arabic for "empty quarter." The very dry, inhospitable interior of the Arabian Peninsula, characterized by large areas of sand and frequent sandstorms.

Sahara Desert. The very large hot desert occupying nearly all of North Africa and adjoining the hot

deserts of Arabia and the Middle East.

Sahel. The southern fringe of the Sahara Desert, characterized by somewhat more abundant rainfall and more vegetation, gradually giving way in the south to grassy plains and then forest.

Scavenger. An animal that feeds primarily on dead animals or their remains.

Somali-Chalbi Desert. A hot, thorny desert on the coast of Ethiopia, Somalia, and Kenya; an eastern extension of the Sahel (see).

Sonoran Desert. The least arid of the hot deserts of North America, located in southern Arizona and northwestern Mexico and having a large variety of desert-adapted plants.

Species. A population of animals or plants capable of freely interbreeding with one another and incapable of interbreeding with individuals of other populations.

Star dune. A large dune with several radiating ridges, the whole forming a star shape and developing in regions with prevailing winds from more than one direction.

Sturt's Stony Desert. A hot desert in the interior of southeastern Australia, characterized by large areas of gibber plains.

Succulent. A plant with thick skin and juicy flesh, capable of storing water and well-adapted to living in desert regions.

Takla Makan Desert. A large cold desert located in the western Chinese province of Sinkiang, lying north of the Tibetan Plateau; the northern part of this desert is sometimes considered a separate division, the Dzungaria Desert (see).

Thar Desert. A hot desert located in northwestern India and eastern Pakistan, often considered an eastern extension of the Sahara and Arabian deserts.

Torpor. A state of reduced metabolic activity into which certain animals enter during periods of aridity or unfavorable temperature, characterized by near cessation of respiration and other vital bodily functions.

Turkestan Desert. A large cold desert located in the southern Soviet Union, east of the Caspian Sea in Turkestan; often regarded as consisting of two parts, the Kara Kum and the Kyzyl Kum (see).

Water pan. A large depression, formed initially by the activity of grazing animals and then widened and deepened by the wind, that catches and holds rainwater.

Water table. The underground water level, below which the soil or rock is saturated with water.

Index

Page numbers in boldface type indicate illustrations